AF581855

Intervertebral Disc Regeneration II

Intervertebral Disc Regeneration II

Editor

Benjamin Gantenbein

MDPI • Basel • Beijing • Wuhan • Barcelona • Belgrade • Manchester • Tokyo • Cluj • Tianjin

Editor
Benjamin Gantenbein
University of Bern
Bern, Switzerland

Editorial Office
MDPI
St. Alban-Anlage 66
4052 Basel, Switzerland

This is a reprint of articles from the Special Issue published online in the open access journal *Applied Sciences* (ISSN 2076-3417) (available at: https://www.mdpi.com/journal/applsci/special_issues/IVD_II).

For citation purposes, cite each article independently as indicated on the article page online and as indicated below:

LastName, A.A.; LastName, B.B.; LastName, C.C. Article Title. *Journal Name* **Year**, *Volume Number*, Page Range.

ISBN 978-3-0365-8198-9 (Hbk)
ISBN 978-3-0365-8199-6 (PDF)

Cover image courtesy of Benjamin Gantenbein

Contents

About the Editor

Benjamin Gantenbein

Benjamin Gantenbein, Prof. Dr., Phil.nat., is an Associated Professor of the Medical Faculty and Group Head of the Bone & Joint Program and the biology lab of Tissue Engineering for Orthopedics & Mechanobiology (TOM) group at the Department for BioMedical Research (DBMR) and the Department of Orthopaedic Surgery and Traumatology, Inselspital, at the Medical Faculty of the University of Bern. His research focuses on intervertebral disc repair using biomaterials and scaffolds, mesenchymal stem cells or a combination thereof. He started his career at the University of Bern in evolutionary biology/phylogenetics. He completed his Master of Science degree and PhD at the Computational and Molecular Population Genetics (CMPG) laboratory at University of Bern. He then received two fellowships (SNF young scientists and one from Marie Curie) to focus on animal molecular evolutionary rates at the Institute of Evolutionary Biology, University of Edinburgh. Finally, he moved to the Genetics and Molecular Ecology Laboratory at Cambridge University, focusing on recombination in animal mitochondria. Before the current assignment, Prof. Gantenbein entered the field of intervertebral disc (IVD) research at the AO Research Institute in Davos, where he acquired experience in modern concepts of tissue engineering and regenerative medicine.

He is active in tissue engineering and stem cell research. He is further involved in bioreactor designs to culture ex vivo bone and joint tissues under physiological conditions. Since 2009, he has been teaching tissue engineering in the Biomedical Engineering Masters course program at the Medical Faculty of the University of Bern.

Preface to "Intervertebral Disc Regeneration II"

This book is devoted to the thriving research on the regeneration of intervertebral discs. It is a colorful mixture of different applied science fields with the shared aim of reducing the number of lower back pain surgeries worldwide. I thank all contributors for submitting their manuscripts to this Special Issue who made this project a success.

Benjamin Gantenbein
Editor

Editorial

New Developments on Growth Factors, Exosomes, and Single Cell RNA-Sequencing for Regeneration of the Intervertebral Disc

Benjamin Gantenbein [1,2]

[1] Tissue Engineering for Orthopaedics & Mechanobiology (TOM), Bone & Joint Program, Department for BioMedical Research (DBMR), Faculty of Medicine, University of Bern, CH-3008 Bern, Switzerland; benjamin.gantenbein@unibe.ch; Tel.: +41-31-632-88-15

[2] Department of Orthopaedic Surgery and Traumatology, Inselspital, Bern University Hospital, Medical Faculty, University of Bern, CH-3010 Bern, Switzerland

1. Introduction

Low back pain (LBP) is the number one cause of disability worldwide, with incidences increasing exponentially [1–3]. A recent study estimates that by the year 2050, an increase of 200 million people is expected, with a current peak at 619 million people [1]. This Special Issue targets the specific niche of finding innovative ways to address the clinical problem of LBP, which is often induced by the prolapse of the spinal column caused by genetic or epigenetic factors. Intervertebral disc (IVD) degeneration is often believed to be the root cause of chronic pain. Future research aims to understand the contribution of metabolic factors such as nutrition, besides other risk factors such as smoking and *Diabetes mellitus* [1]. This Special Issue provides a heterogeneous snapshot of recent applied research on IVD and LBP, ranging from cell biology studies to artificial intelligence in diagnostics. In the following subsections, I provide a short overview and summarize the seven articles' main findings.

2. Wet Laboratory Studies in the Second Special Issue

Two original studies focused on the single-cell transcriptomics to characterize phenotypes in rats in an IVD degeneration model [4] or performed a pathway study of ~100 genes using qPCR gene arrays using total RNA extractions from human donors suffering from idiopathic skeletal hyperostosis (DISH). The first study established an in vivo IVD degeneration model in 8-week-old Sprague Dawley rats that underwent surgery for retroperitoneal exposure using a 27 Gauge needle of the L4-L6 lumbar spine. Rohanifar et al. (2022) [4] then allowed the rats to recover for two and eight weeks postoperatively. Then, they digested the single cells from the tissue using mild digestion protocols, extracted and sequenced the total RNA and compared the next-generation sequencing (NGS) data relative to untreated controls. Rohanifar et al. (2022) confirmed that the nucleus pulposus (NP) mainly expressed key markers such as CD24 [5] as well as aggrecan and collagen type 2 [6–8]. In the outer part of the IVD, the cells mainly expressed collagen type 1, as previously identified in rats [6–8] but also in other species such as bovine and human [9]. Furthermore, their data allowed us to distinguish IVD cells from lymphoid, endothelial and myeloid cells [4]. The needle-punctured groups subsequently had a significantly higher amount of myeloid cells and lymphocytes than controls. These data are logical since needle puncture causes inflammation [4]. The second transcriptome study is on DISH patients. DISH is also known as Forestier's disease [10–13]. To the best of my knowledge, the study by Gantenbein et al. (2021) [14] is the first to elucidate on the possible phenotypic changes and deregulations of DISH cells compared to the phenotype of IVD cells isolated from trauma patients. This comparison has its limitations, of course, as trauma cells are not necessarily "healthy" cells. However, it most likely that these cells are still in a better state than cells

Citation: Gantenbein, B. New Developments on Growth Factors, Exosomes, and Single Cell RNA-Sequencing for Regeneration of the Intervertebral Disc. *Appl. Sci.* **2023**, *13*, 7346. https://doi.org/10.3390/app13137346

Received: 12 June 2023
Accepted: 19 June 2023
Published: 21 June 2023

isolated from degenerated IVDs. Nevertheless, by comparing the expression levels of bone morphogenic protein pathway cytokines and their inhibitors, i.e., Gremlin, Noggin and Chordin, my group found that these were dys-regulated [15], although these changes were not significant. However, DISH-IVD, in contrast to IVD obtained from trauma, showed a significant up-regulation of early growth response 2 (EGR2) interleukin 6 (IL6), and insulin-like growth factor 1 (IGF1) tended to be up-regulated [14]. IGF1 has been proposed by other authors to be a possible marker in serum samples of DISH patients [16,17].

The third study in the Special Issue is that of Bischof et al. (2021) [18], who focused on cell culture and tissue-specific progenitor cells. They presented a study on the optimization of culture conditions on the so-called nucleus pulposus progenitor cells (NPPC). In this study, primary IVD cells were isolated from human disc tissue with a mild digestion protocol [19,20]. After reaching confluence in monolayer recovery, these mixed IVD cells, mainly from the nucleus pulposus (NP), were then trypsinized and sorted with a surface marker named Tie2 (or CD202b), which stands for angiopoeitin receptor-1. The Tie2+ cells and the Tie2- cells were then further cultured in normoxic and hypoxic (i.e., 2%) conditions in presence of Angiopoetin-1 (Ang-1) or Ang-2 at increasing doses [21]. However, the results of the study [18] did not produce the expected results, namely, that Tie2+ were stimulated and would proliferate faster compared to Tie2− cells. Despite this, it seemed very clear that hypoxia, i.e., at 2% oxygen level, was the most important factor for higher cellular metabolic activity. Their conclusion that hypoxia is beneficial for these NPPC is in agreement with other studies performed by the team of Daisuke Sakai from Tokai University School of Medicine [22,23].

3. Clinical/Radiological Studies in the Second Special Issue

There are two radiological studies published in this Special Issue. Firstly, Landauer and Trieb (2022) [24] provide radiological evidence that the lumbosacral transitional vertebrae (LSTV) are valid as a model for IVD regeneration. They scanned 1482 patients radiologically, and their LSTV were then classified according to Castellvi classification type II–IV [24]. Additionally, magnetic resonance scans (MRIs) were also obtained from selected patients. The authors concluded that the reduced or absent mobility in the LSTV segments led to an overload of the adjacent segments in these patients.

The second study, by Kim et al. (2022) [25], is on the usage of natural language processing (NLP), which is defined as understanding, analyzing, and extracting meaningful information from text (natural language) by computer science [26]. This research targets a highly significant area of research, which is "big data". It is obvious that artificial intelligence (AI) will be necessary to make full use of all the available clinical data and to help surgeons to take decisions with the assistance of fast data processing. In this approach, the authors tested their NLP pipeline on a balanced sample of 300 X-ray, 300 CT, and 300 MRI reports. When evaluating their NLP model performances, four parameters—recall, precision, accuracy, and a so-called "F1" score (the harmonic mean of precision and recall [27])—were greater than 0.9 for all 23 radiologic findings.

4. A Review on Secretomes of the IVD

Extracellular vesicles (EVs) have long attracted the attention of the regenerative community. The importance of this topic is underlined by a current "wave" of comparable articles that also focus on the usage of EVs to regenerate the IVD [28–31]. This is not altogether surprising as regulatory hurdles toward proving non-toxicity and patient safety have recently been introduced by authorities such as the Federal Drug and food Agency (FDA) and the label of the Conformité Européene (CE) [32]. This applies in particular to cellular applications. Thus, secretomes, or so-called conditioned media, have been the focus of recent IVD-related research [29,33]. The review by Tilotta et al. (2023) provides a valuable insight into the field of EVs and a summary of their characterization [29].

5. Conclusions

Overall, this Special Issue offers a good insight into the heterogeneity of IVD research and the recent findings not only from clinics, but also from biologists and engineers. I hope that this Special Issue will give scientists an overview of this highly translational and applied fast growing research field. There is still yet further research to come to help to find possible "cures" for affected patients. With the recent prognosis by Ferreira et al. (2023) [1] warning of an increase of one-third more LBP patients in the next 50 years, the scientific community is urged to find better treatment options and also especially early diagnostic tools to foresee critical cases to come.

Acknowledgments: B. Gantenbein wishes to acknowledge his funding sources, namely, the Marie Skłodowska Curie International Training Network (ITN) "disc4all" (https://disc4all.upf.edu, accessed on 6 June 2023) grant agreement #955735 (https://cordis.europa.eu/project/id/955735, accessed on 6 June 2023), the iPspine project, and the Swiss National Science Foundation projects #310030E_192674/1 (https://data.snf.ch/grants/grant/192674, accessed on 6 June 2023) and project #40B2-0_211510/1 (https://data.snf.ch/grants/grant/211510, accessed on 6 June 2023).

Conflicts of Interest: The author declares no conflict of interest.

References

1. Ferreira, M.L.; de Luca, K.; Haile, L.M.; Steinmetz, J.D.; Culbreth, G.T.; Cross, M.; Kopec, J.A.; Ferreira, P.H.; Blyth, F.M.; Buchbinder, R.; et al. Global, regional, and national burden of low back pain, 1990–2020, its attributable risk factors, and projections to 2050: A systematic analysis of the Global Burden of Disease Study 2021. *Lancet Rheumatol.* **2023**, *5*, e316–e329. [CrossRef]
2. GBD 2017 Disease and Injury Incidence and Prevalence Collaborators. Global, regional, and national incidence, prevalence, and years lived with disability for 354 diseases and injuries for 195 countries and territories, 1990–2017: A systematic analysis for the Global Burden of Disease Study 2017. *Lancet* **2018**, *392*, 1789–1858. [CrossRef] [PubMed]
3. Rivera-Tavarez, C.E. Can We Increase Our Health Span? *Phys. Med. Rehabil. Clin. N. Am.* **2017**, *28*, 681–692. [CrossRef] [PubMed]
4. Rohanifar, M.; Clayton, S.W.; Easson, G.W.; Patil, D.S.; Lee, F.; Jing, L.; Barcellona, M.N.; Speer, J.E.; Stivers, J.J.; Tang, S.Y.; et al. Single Cell RNA-Sequence Analyses Reveal Uniquely Expressed Genes and Heterogeneous Immune Cell Involvement in the Rat Model of Intervertebral Disc Degeneration. *Appl. Sci.* **2022**, *12*, 8244. [CrossRef]
5. Liu, Z.; Zheng, Z.; Qi, J.; Wang, J.; Zhou, Q.; Hu, F.; Liang, J.; Li, C.; Zhang, W.; Zhang, X. CD24 identifies nucleus pulposus progenitors/notochordal cells for disc regeneration. *J. Biol. Eng.* **2018**, *12*, 35. [CrossRef]
6. Minogue, B.M.; Richardson, S.M.; Zeef, L.A.; Freemont, J.; Hoyland, J.A. Transcriptional profiling of bovine intervertebral disc cells: Implications for identification of normal and degenerate human intervertebral disc cell phenotypes. *Thromb. Haemost.* **2010**, *12*, R22. [CrossRef]
7. Shapiro, I.M.; Risbud, M.V. Transcriptional profiling of the nucleus pulposus: Say yes to notochord. *Thromb. Haemost.* **2010**, *12*, 117. [CrossRef]
8. Rutges, J.; Creemers, L.; Dhert, W.; Milz, S.; Sakai, D.; Mochida, J.; Alini, M.; Grad, S. Variations in gene and protein expression in human nucleus pulposus in comparison with annulus fibrosus and cartilage cells: Potential associations with aging and degeneration. *Osteoarthr. Cartil.* **2010**, *18*, 416–423. [CrossRef] [PubMed]
9. Calió, M.; Gantenbein, B.; Egli, M.; Poveda, L.; Ille, F. The cellular composition of bovine coccygeal intervertebral discs: A comprehensive single-cell RNAseq analysis. *Int. J. Mol. Sci.* **2021**, *22*, 4917. [CrossRef]
10. Resnick, D.; Shapiro, R.F.; Wiesner, K.B.; Niwayama, G.; Utsinger, P.D.; Shaul, S.R. Diffuse idiopathic skeletal hyperostosis (DISH) [ankylosing hyperostosis of forestier and Rotes-Querol]. *Semin. Arthritis Rheum.* **1978**, *7*, 153–187. [CrossRef]
11. Resnick, D.; Shaul, S.R.; Robins, J.M. Diffuse Idiopathic Skeletal Hyperostosis (DISH): Forestier's Disease with Extraspinal Manifestations. *Radiology* **1975**, *115*, 513–524. [CrossRef] [PubMed]
12. Forestier, J.; Rotes-Querol, J. Senile Ankylosing Hyperostosis of the Spine. *Ann. Rheum. Dis.* **1950**, *9*, 321–330. [CrossRef] [PubMed]
13. Mader, R.; Verlaan, J.-J.; Buskila, D. Diffuse idiopathic skeletal hyperostosis: Clinical features and pathogenic mechanisms. *Nat. Rev. Rheumatol.* **2013**, *9*, 741–750. [CrossRef]
14. Gantenbein, B.; May, R.D.; Bermudez-Lekerika, P.; Oswald, K.A.C.; Benneker, L.M.; Albers, C.E. EGR2, IGF1 and IL6 Expression Are Elevated in the Intervertebral Disc of Patients Suffering from Diffuse Idiopathic Skeletal Hyperostosis (DISH) Compared to Degenerative or Trauma Discs. *Appl. Sci.* **2021**, *11*, 4072. [CrossRef]
15. May, R.D.; Frauchiger, D.A.; Albers, C.E.; Tekari, A.; Benneker, L.M.; Klenke, F.M.; Hofstetter, W.; Gantenbein, B. Application of Cytokines of the Bone Morphogenetic Protein (BMP) Family in Spinal Fusion—Effects on the Bone, Intervertebral Disc and Mesenchymal Stromal Cells. *Curr. Stem Cell Res. Ther.* **2019**, *14*, 618–643. [CrossRef]

16. Denko, C.W.; Malemud, C.J. Body mass index and blood glucose: Correlations with serum insulin, growth hormone, and insulin-like growth factor-1 levels in patients with diffuse idiopathic skeletal hyperostosis (DISH). *Rheumatol. Int.* **2005**, *26*, 292–297. [CrossRef]
17. Denko, C.W.; Boja, B.; Moskowitz, R.W. Growth factors, insulin-like growth factor-1 and growth hormone, in synovial fluid and serum of patients with rheumatic disorders. *Osteoarthr. Cartil.* **1996**, *4*, 245–249. [CrossRef]
18. Bischof, M.C.; Häckel, S.; Oberli, A.; Croft, A.S.; Oswald, K.A.C.; Albers, C.E.; Gantenbein, B.; Guerrero, J. Influence of Angiopoietin Treatment with Hypoxia and Normoxia on Human Intervertebral Disc Progenitor Cell's Proliferation, Metabolic Activity, and Phenotype. *Appl. Sci.* **2021**, *11*, 7144. [CrossRef]
19. Sakai, D.; Schol, J.; Bach, F.C.; Tekari, A.; Sagawa, N.; Nakamura, Y.; Chan, S.C.; Nakai, T.; Creemers, L.B.; Frauchiger, D.A.; et al. Successful fishing for nucleus pulposus progenitor cells of the intervertebral disc across species. *JOR Spine* **2018**, *1*, e1018. [CrossRef]
20. Frauchiger, D.A.; Tekari, A.; May, R.D.; Džafo, E.; Chan, S.C.; Stoyanov, J.; Bertolo, A.; Zhang, X.; Guerrero, J.; Sakai, D.; et al. Fluorescence-Activated Cell Sorting Is More Potent to Fish Intervertebral Disk Progenitor Cells Than Magnetic and Beads-Based Methods. *Tissue Eng. Part C Methods* **2019**, *25*, 571–580. [CrossRef] [PubMed]
21. Sakai, D.; Nakamura, Y.; Nakai, T.; Mishima, T.; Kato, S.; Grad, S.; Alini, M.; Risbud, M.V.; Chan, D.; Cheah, K.S.; et al. Exhaustion of nucleus pulposus progenitor cells with ageing and degeneration of the intervertebral disc. *Nat. Commun.* **2012**, *3*, 1264. [CrossRef] [PubMed]
22. Sako, K.; Sakai, D.; Nakamura, Y.; Matsushita, E.; Schol, J.; Warita, T.; Horikita, N.; Sato, M.; Watanabe, M. Optimization of Spheroid Colony Culture and Cryopreservation of Nucleus Pulposus Cells for the Development of Intervertebral Disc Regenerative Therapeutics. *Appl. Sci.* **2021**, *11*, 3309. [CrossRef]
23. Sako, K.; Sakai, D.; Nakamura, Y.; Schol, J.; Matsushita, E.; Warita, T.; Horikita, N.; Sato, M.; Watanabe, M. Effect of Whole Tissue Culture and Basic Fibroblast Growth Factor on Maintenance of Tie2 Molecule Expression in Human Nucleus Pulposus Cells. *Int. J. Mol. Sci.* **2021**, *22*, 4723. [CrossRef] [PubMed]
24. Landauer, F.; Trieb, K. Diagnostic Limitations and Aspects of the Lumbosacral Transitional Vertebrae (LSTV). *Appl. Sci.* **2022**, *12*, 10830. [CrossRef]
25. Kim, Y.; Song, C.; Song, G.; Kim, S.B.; Han, H.-W.; Han, I. Using Natural Language Processing to Identify Low Back Pain in Imaging Reports. *Appl. Sci.* **2022**, *12*, 12521. [CrossRef]
26. Bacco, L.; Russo, F.; Ambrosio, L.; D'antoni, F.; Vollero, L.; Vadalà, G.; Dell'orletta, F.; Merone, M.; Papalia, R.; Denaro, V. Natural language processing in low back pain and spine diseases: A systematic review. *Front. Surg.* **2022**, *9*, 957085. [CrossRef] [PubMed]
27. Syed, K.; Sleeman, W.; Hagan, M.; Palta, J.; Kapoor, R.; Ghosh, P. Automatic Incident Triage in Radiation Oncology Incident Learning System. *Healthcare* **2020**, *8*, 272. [CrossRef]
28. Teo, K.Y.W.; Tan, R.; Wong, K.L.; Hey, D.H.W.; Hui, J.H.P.; Toh, W.S. Small extracellular vesicles from mesenchymal stromal cells: The next therapeutic paradigm for musculoskeletal disorders. *Cytotherapy* **2023**. [CrossRef] [PubMed]
29. Tilotta, V.; Vadalà, G.; Ambrosio, L.; Cicione, C.; Di Giacomo, G.; Russo, F.; Papalia, R.; Denaro, V. Mesenchymal stem cell-derived secretome enhances nucleus pulposus cell metabolism and modulates extracellular matrix gene expression in vitro. *Front. Bioeng. Biotechnol.* **2023**, *11*, 1152207. [CrossRef]
30. Yu, H.; Wang, H.; Lin, S. Exosome-mediated repair of intervertebral disc degeneration: The potential role of miRNAs. *Curr. Stem Cell Res. Ther.* **2023**. [CrossRef]
31. Zhou, H.-Y.; Hu, Y.-C.; Zhang, X.-B.; Lin, M.-Q.; Cong, M.-X.; Chen, X.-Y.; Zhang, R.-H.; Yu, D.-C.; Gao, X.-D.; Guo, T.-W. Nanoscale treatment of intervertebral disc deneration: Mesenchymal stem cell exosome transplantation. *Curr. Stem. Cell Res. Ther.* **2022**, *18*, 163–173. [CrossRef] [PubMed]
32. Evans, C.; Ghivizzani, S.; Robbins, P. Orthopedic gene therapy-lost in translation? *J. Cell. Physiol.* **2011**, *227*, 416–420. [CrossRef] [PubMed]
33. Zhao, X.; Xu, B.; Duan, W.; Chang, L.; Tan, R.; Sun, Z.; Ye, Z.; Md, D.P. Insights into Exosome in the Intervertebral Disc: Emerging Role for Disc Homeostasis and Normal Function. *Int. J. Med. Sci.* **2022**, *19*, 1695–1705. [CrossRef] [PubMed]

Review

Mesenchymal Stem Cell-Derived Exosomes: The New Frontier for the Treatment of Intervertebral Disc Degeneration

Veronica Tilotta, Gianluca Vadalà *, Luca Ambrosio, Fabrizio Russo, Claudia Cicione, Giuseppina Di Giacomo, Rocco Papalia and Vincenzo Denaro

Laboratory for Regenerative Orthopaedics, Department of Orthopaedic and Trauma Surgery, Campus BioMedico University of Rome, Via Alvaro del Portillo 200, 00128 Rome, Italy; v.tilotta@unicampus.it (V.T.); l.ambrosio@unicampus.it (L.A.); fabrizio.russo@unicampus.it (F.R.); c.cicione@unicampus.it (C.C.); g.digiacomo@unicampus.it (G.D.G.); r.papalia@unicampus.it (R.P.); denaro@unicampus.it (V.D.)

* Correspondence: g.vadala@unicampus.it

Abstract: Low back pain (LBP) is one of the most frequent symptoms associated with intervertebral disc degeneration (IDD) and affects more than 80% of the population, with strong psychosocial and economic impacts. The main cause of IDD is a reduction in the proteoglycan content within the nucleus pulposus (NP), eventually leading to the loss of disc hydration, microarchitecture, biochemical and mechanical properties. The use of mesenchymal stem cells (MSCs) has recently arisen as a promising therapy for IDD. According to numerous reports, MSCs mediate their regenerative and immunomodulatory effects mainly through paracrine mechanisms. Recent studies have suggested that extracellular vesicles (EVs) extracted from MSCs may be a promising alternative to cell therapy in regenerative medicine. EVs, including exosomes and microvesicles, are secreted by almost all cell types and have a fundamental role in intercellular communication. Early results have demonstrated the therapeutic potential of MSCs-derived EVs for the treatment of IDD through the promotion of tissue regeneration, cell proliferation, reduction in apoptosis and modulation of the inflammatory response. The aim of this review is to focus on the biological properties, function, and regulatory properties of different signaling pathways of MSCs-derived exosomes, highlighting their potential applicability as an alternative cell-free therapy for IDD.

Keywords: extracellular vesicles; exosomes; mesenchymal stem cells; miRNA; low back pain; intervertebral disc degeneration; regenerative medicine

Citation: Tilotta, V.; Vadalà, G.; Ambrosio, L.; Russo, F.; Cicione, C.; Di Giacomo, G.; Papalia, R.; Denaro, V. Mesenchymal Stem Cell-Derived Exosomes: The New Frontier for the Treatment of Intervertebral Disc Degeneration. *Appl. Sci.* **2021**, *11*, 11222. https://doi.org/10.3390/app112311222

Academic Editor: Benjamin Gantenbein

Received: 18 October 2021
Accepted: 25 November 2021
Published: 26 November 2021

1. Introduction

Low back pain (LBP) is one of the most common musculoskeletal symptoms affecting up to 80% of adults aged between 40–80 years. Chronic LBP is one of the main causes of disability and loss of work ability with a strong impact on patients' quality of life; additionally, it poses a significant socioeconomic burden on public health worldwide [1]. According to a recent global report on the prevalence of LBP, it has been estimated that two thirds of adults are afflicted with LBP at least once in their lives. In addition, the financial burden of LBP in the United States is estimated to exceed $100 billion per year, depending on the inevitable effect of changes in work status, prolonged hospitalization, and increased outpatient visits [2–4].

The majority of LBP episodes are associated with intervertebral disc degeneration (IDD). The intervertebral disc (IVD) acts as a shock absorber between the vertebrae while allowing for flexion–extension, lateral flexion, and rotation movements. The IVD is a complex structure connecting two adjacent vertebrae and is composed of the annulus fibrosus (AF), nucleus pulposus (NP) and cartilaginous endplates (CEP) [5]. The AF is a fibro-cartilaginous ring encircling the NP and it consists of concentric lamellae of type I collagen fibers associated with an interlamellar matrix, in which scattered cells with fibroblast-like morphology and phenotype are present [6]. The NP presents a gelatinous extracellular matrix (ECM) rich in proteoglycans (mainly aggrecan), type II collagen and other proteins such as elastin and fibronectin. The high level of water content helps maintain disc height and supports the functions of the IVD [7]. The NP cell population has a notochordal origin: notochordal cells (NCs) act as progenitor cells and play an important role in stimulating ECM synthesis by NP cells (NPCs), although in adult human NP tissue the number of NCs diminishes drastically [8]. To preserve IVD architectural features, it is essential to maintain disc cell metabolism and physiological biomechanics, but the avascular and relatively acellular environment of the NP leads to a limited regenerative capacity against stressful and degenerative stimuli. IDD is an age-related disease and is mainly linked to loss of NP hydration and reduction in the proteoglycan content [9]. Current approaches to treat IDD are based on conservative or surgical procedures that aim to relieve the symptoms; however, they are not able to change the natural history of the disease [10].

Recent advances in the understanding of IVD biology have led to an increased interest in the development of novel biological treatments. Growth factors, gene therapy, stem cell transplantation and tissue engineering might support intervertebral disc regeneration [11–16]. The transplantation of mesenchymal stem cells (MSCs), including bone marrow-derived (BM-MSCs) and adipose tissue (Ad-MSCs), has been evaluated in different clinical trials. Although encouraging results have been achieved, there are still several impediments related to the survival and differentiation of transplanted MSCs due to the particularly hostile microenvironment of degenerated IVDs [17]. Moreover, costs related to cell expansion under Good Manufacturing Practice (GMP) conditions and to the quality control guidelines for Advanced Therapy Medical Products (ATMP) limits the large-scale application of a cell therapy strategy.

In the last decade, the interest in extracellular vesicles (EVs) and their therapeutic potential has grown regarding different applications in several fields, including oncology, neurology, cardiology, and orthopedics [18,19]. EVs are endogenous nanoparticles released by different cell types and are involved in various physiological mechanisms including apoptosis, tissue regeneration, inflammation, ECM synthesis, cell proliferation and immunomodulation [20]. Based on morphology, biogenesis, size, and content, EVs are generally distinguished in apoptotic bodies, microvesicles and exosomes [21]. One of the contemporary challenges in regenerative medicine is the development of new biology-based therapeutic strategies employing EVs. Indeed, MSCs-derived exosomes have recently demonstrated their potential benefits in preclinical models of cardiovascular diseases, graft-versus-host disease and osteoarthritis [19], while their role in IDD still remains unclear [22].

The aim of this review is to focus on the biological properties, function, and regulatory properties of different signaling pathways of MSCs-derived exosomes, highlighting their potential applicability as an alternative cell-free therapy for IDD.

2. Intervertebral Disc Degeneration

IDD is an age-related chronic and progressive process, characterized by a gradual loss of proteoglycans and water content in the NP with a reduction in the disc's capacity to withstand compressive loads [17].

The IVD is an avascular organ, and the supply of nutrition occurs by diffusion from its surrounding vasculature through the CEPs [23]. However, IVD aging and mechanical overload progressively foster endplate calcification and capillary obliteration, thus reducing nutrient delivery to the IVD itself [24]. Due to the lack of blood flow, a hypoxic microenvironment is consequently established with higher O_2 concentrations at the surface of the AF (19.5%) and a reduction towards the center of the NP (0.65%). However, low oxygen concentration has not been shown to be detrimental for NPC viability [25]. Another factor involved in making the IVD microenvironment hostile is the low glucose concentration, which ranges from about 5 mM at the periphery to about 0.8 mM at the center of the IVD [26].

pH is another crucial parameter in the IVD microenvironment. Local production of lactic acid by IVD maintains a pH between 2 and 6, with significant effects on cell survival and matrix synthesis, as well as on the expression of proinflammatory cytokines and pain-related factors [27]. The main change in NP during IDD is in the imbalance between anabolic factors, such as insulin-like growth factor 1 (IGF-1), growth differentiation factor 6 (GDF-6), transforming growth factor β (TGF-β), bone morphogenetic proteins (BMPs) and catabolic enzymes, such as matrix metalloproteinases (MMPs) and a disintegrin and metalloproteinase with thrombospondin motifs (ADAMTS) [28].

Aggrecan, the most abundant proteoglycan in the NP matrix, binds to sulphated glycosaminoglycans (GAGs, e.g., keratan and chondroitin sulfate), and consequently retains hydration and osmotic pressure within the IVD. Negatively charged GAGs regulate the ion balance of the ECM [29]. In a healthy state, osmolarity can range from 430 mOsm/L (isosmotic pressure) to about 496 mOsm/L (hyperosmotic pressure) [30]. In degenerative conditions, due to the progressive loss of proteoglycans, the osmolarity of the IVD decreases until it reaches values of about 300 mOsm/L [31]. Tao et al. reported that high osmolarity can not only decrease the viability and proliferation of disc cells but can also influence the expression levels of collagen type II, SOX-9 and aggrecan in progenitor NPCs [32].

Excessive mechanical loading may be another crucial factor that compromises the biomechanical characteristics of IVD. Static compression stress induces apoptosis, senescence, mitochondrial damage, and matrix catabolism due to a reduction in collagen and aggrecan expression with an increase in MMPs [33]. On the contrary, dynamic compressive loading often triggers an anabolic response and improves the transport of nutrients and growth factors within the NP [34].

The inflammatory response may also contribute to the onset and progression of the degenerative process [35]. The nutritional decrease and accumulation of degraded matrix products promote macrophage-mediated production of pro-inflammatory cytokines such as interleukin (IL)-1β and tumor necrosis factor-α (TNF-α) through activation of the nuclear factor kappa-light-chain-enhancer of activated B cells (NF-kB) pathway [36]. IL-1β and TNF-α are involved in the progression and initiation of IDD by regulating inflammatory response, IVD cell proliferation, senescence, apoptosis, pyroptosis, autophagy, ECM destruction, and oxidative stress [37]. TNF-α is able to amplify the inflammatory response by stimulating IVD cells to synthesize additional cytokines, including IL-6, IL-8, IL-17, IL-1β and numerous chemokines [38]. IL-1β is expressed at higher levels in degenerated IVDs with an increased production of genes and proteins for the IL-1 receptor. Therefore, although both cytokines are involved in the pathogenesis of IDD, IL-1β may play a more significant role, and could thus be considered an ideal therapeutic target [39]. Tissue aging increases oxidative stress, mitochondrial dysfunction, DNA damage, cell senescence and apoptosis, thus compromising ECM homeostasis [40]. Besides, the viability of NPCs has an important impact on ECM composition. The declining number of cells, their phenotype modification and the inability to withhold compression stresses lead to the deterioration of the AF lamellar architecture and the reduction in ECM synthesis [41,42]. The low metabolic activity of cells and the avascular nature of IVD may explain the inability to self-repair following injury and degeneration [43]. However, recent research has reported the presence of progenitor stem cells in the NP, which may be physiologically responsible, to a certain extent, for endogenous repair of intrinsic insults to IVD integrity.

The natural history of IDD may progress to disc herniation, degenerative spondylolisthesis, spinal instability, and stenosis associated with neurological symptoms such as radiculopathy and/or myelopathy, and surgical intervention is often required [44]. However, the available treatment approaches are not able to arrest IDD; they are only able to relieve accompanying symptoms. Therefore, strenuous research efforts are being made to develop effective strategies that can achieve IVD regeneration. Therefore, the potential use of EVs appears to be a promising therapeutic tool that may support endogenous repair and re-establish the proteoglycan content, thus improving disc hydration and biomechanics.

3. MSC Interaction with NP Cells

Human MSCs are multipotent cells present in almost all tissues such as bone marrow, adipose tissue, skin, lung, synovial membrane, dental pulp, nasal olfactory mucosa, muscle, periosteum, corneal limbus, peripheral blood, endometrial and menstrual blood, uterine cervix, and in fetal/neonatal tissues [9]. Stem cell transplantation represents a promising approach to restore the homeostasis of the degenerated environment of IVD due to the ability of the stem cells to differentiate towards NP-like cells and their immunomodulatory and anti-catabolic effects [14]. In recent decades, several preclinical studies have evaluated the interaction between MSCs and NPCs in vitro [44–46].

Cunha et al. tested the systemic delivery of BM-MSCs in a rat IVD and reported a diminished hypoxic response, risk of herniation and synthesis of cytokines [45]. In an experimental study, Wangler et al. described the homing of MSCs as a potential strategy to prevent the onset of the degenerative cascade in IVDs. They have also evaluated the effect of human MSCs homing in bovine and human disc cells expressing Tie2 (also known as angiopoietin-1 receptor TEK tyrosine kinase). $Tie2^{+}$ cells represent a progenitor cell population with discogenic differentiation potential. The results showed a proliferative response and reduction in dead cells in both tissues [46].

Our group was one of the first to report the crosstalk mechanism between MSCs and NPCs. In a 3D culture system, where a short distance may improve communication among cells, a paracrine effect of MSCs on NPCs collected from a degenerative disc was demonstrated. Gene expression analysis on both cell types showed that MSCs exerted a trophic effect on NPCs while acquiring a chondrogenic phenotype due to the communication with NPCs [47]. In another similar in vitro study, we highlighted the significant increase in GAG content and proliferation of NPCs when co-cultured with MSCs [48].

In the last decade, intradiscal stem cell therapy for IDD has provoked increasing interest in the field. Since cell isolation and implantation procedures can be safely performed in a sterile environment and with low immunogenicity, both autologous and allogeneic transplants can be used [49]. In a pilot study by Orozco et al., 10 patients with chronic LBP were injected with autologous BM-MSCs for the treatment of lumbar disc degeneration. Three months after surgery, LBP and disability were significantly reduced. Despite the small number of patients, this study demonstrated that intradiscal BM-MSC transplantation was a safe procedure with a promising role in the treatment of IDD [50]. Similarly, a randomized controlled trial including 24 patients with lumbar IDD reported a significant reduction in pain in the lumbar region and reduced disability one year after BM-MSC allogeneic transplant, whereas the control group, who received a sham infiltration within the paravertebral musculature did not show any improvement [51].

Earlier research has proved that MSCs secrete bioactive molecules and trophic factors such as soluble proteins, nucleic acids, lipids and EVs, and highlights their anti-apoptotic, immunomodulatory, angiogenetic, anti-fibrotic functions and differentiation capacity, probably mediated by paracrine mechanisms [52–54]. The products released by MSCs in vitro and in vivo constitute their secretome [55]. In a bovine ex vivo model, Teixeira et al. investigated the immunomodulatory paracrine effect of BM-MSCs in the proinflammatory IDD microenvironment. While MSCs viability was not influenced, cell migration increased, collagen type II and aggrecan levels remained stable while inflammatory cytokines IL-6, IL-8 and TNF-α were downregulated by MSCs in IL-1β-stimulated IVDs [56]. Furthermore, MSCs have been shown to migrate from their niches to damaged sites in response to injury, immune or physicochemical signals. This physiological mechanism of MSCs recruitment or homing may play a role in promoting the repair of different musculoskeletal tissues [57].

Interestingly, recent studies investigated the cross-talk between MSCs and NPCs through trophic factors [58]. Shim et al. investigated paracrine interactions in a co-culture system by isolating MSCs, AF and NPCs from the same donor. The characterization revealed an increase in mRNA levels expression of ECM genes (SOX9, COL2A1 and ACAN), the downregulation of MMPs and ADAMTS, as well as the inhibition of the production of pro-inflammatory factors compared to monocultures [59]. Li et al. also employed a co-culture system to understand the influence of BM-MSCs on TNF-α-induced NPC degeneration and reported an increase in cell proliferation and type II collagen production, a decrease in MMP-9 expression, and reduced TGFβ/NF-κB signaling in pro-senescent NPCs. Premature senescence and aging of NPCs were restored in the co-culture system [60]. Even though the regenerative, anabolic, and inflammatory-modulating properties of MSCs have been demonstrated in several investigations, many aspects are still unclear.

The harsh microenvironment with the absence of vasculature, hypoxia, low glucose concentration, acidity, hyperosmolarity and continuous mechanical stimulation plays a crucial role for cell survival and their regenerative potential [61]. Furthermore, MSC implantation may further impair the balance between the nutrient supply and demand of resident disc cells, thus causing cell death. Miguélez-Rivera et al. investigated the effectiveness of a conditioned medium (CM) derived from the culture of MSCs in a rat IDD model. MSC-derived CM has shown immunomodulatory and anti-inflammatory properties in a detrimental environment by down-regulating the expression levels of pro-inflammatory cytokines such as IL-1β, IL-6, IL-17 and TNF [62].

Although MSC-based therapy may be considered an attractive and a safe option, there are still unsolved issues. In addition to the abovementioned shortcomings, MSCs require a long and expensive process for in vitro expansion according to the minimum standard to guarantee controlled conditions in production processes and clinical procedures [63].

4. Extracellular Vesicles

EVs were discovered in 1987 by Johnstone et al. during the in vitro maturation of reticulocytes in culture [64]. Secreted EVs are a group of cell-derived membranous structures with biological potential in numerous human diseases (e.g., cancer, neurodegenerative and musculoskeletal disorders). Collection, pre-processing, separation, concentration, and characterization of EVs have been widely studied. In 2018, the International Society for Extracellular Vesicles (ISEV) proposed the Minimal Information for Studies of Extracellular Vesicles (MISEV) as an update of general knowledges presented in MISEV 2014 guidelines to allow the reproducibility of results among investigators [65].

4.1. Size and Morphology

The determining factors that discriminate the different classes of EVs are size, morphology, origin and content (Table 1) [66].

Table 1. General features of exosomes, microvesicles and apoptotic bodies.

	Exosomes	Microvesicles	Apoptotic Bodies
Size	30–100 nm	50–1000 nm	800–5000 nm
Morphology	Cup-shaped	Spherical structures	Heterogeneous
Content	Coding RNA, noncoding RNA, proteins, lipids, DNA	RNA, proteins, lipids, cytosol	Proteins, lipids, DNA, RNA, cytosol
Biogenesis	By exocytosis in which ILVs within the lumen of MVEs fuse with the plasma membrane to release ILVs	By outward budding and fission of the plasma membrane and the sub-sequent release of vesicles into the extracellular space	Outward blebbing of the cell membrane

ILVs = intraluminal vesicles; MVEs = multivesicular endosomes.

It is possible to distinguish three main EV categories: apoptotic bodies, microvesicles and exosomes (Figure 1) [67].

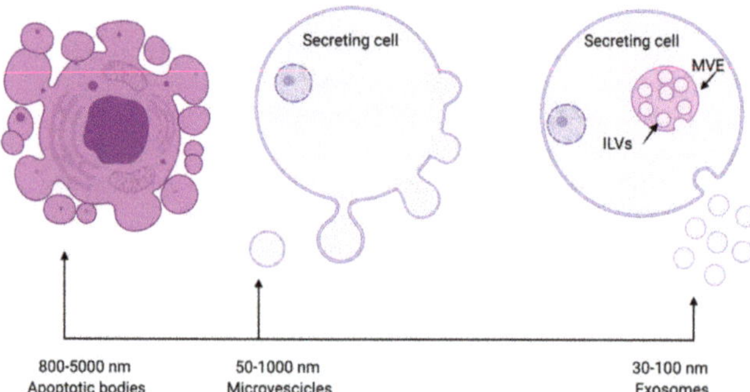

Figure 1. Extracellular vesicles distinguished by size: apoptotic bodies (800–5000 nm), microvesicles (50–1000 nm) and exosomes (30–100 nm). Created with BioRender.com (accessed on 10 October 2021).

Apoptotic bodies have a size range of 800–5000 nm with a heterogeneous morphology. They also include smaller vesicles known as apoptotic cell-derived microparticles or apoptotic blebs released by apoptotic cells [66].

Microvesicles were originally described as "platelet dust", as they were first isolated from platelets [68]. They have a diameter of 50–1000 nm, a spherical structure, and are expelled through exocytosis of the plasma membrane [69].

The last class of vesicles are smaller than those mentioned above, have a size range of 30–100 nm, a cup-shaped morphology and are called exosomes [70]. However, some studies have shown that the cup-shaped morphology could be an experimental artefact and the overlapping range of size may lead to unclear terminology [71]. Further isolation and characterization protocols are expected to establish an appropriate classification and nomenclature [72].

4.2. Content and Biogenesis

Cell type and the physiological or pathological state of the parent cell may influence its EV content [73]. To date, the interest in EVs has grown exponentially due to their abundant presence in body fluids, their wide range of regulatory functions and their capacity as vehicles for DNA, mRNAs, miRNAs, proteins, and lipids involved in intercellular communication (Figure 2) [74].

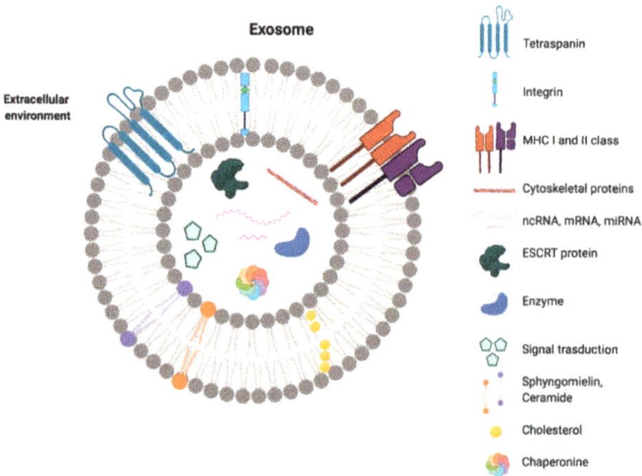

Figure 2. Exosomal composition in terms of bioactive molecules: lipids, proteins, and nucleic acids. Created with BioRender.com (accessed on 10 October 2021).

The protein content of EVs is often used as a marker to recognize EV sub-populations. Exosomes are rich in endosome-associated proteins (e.g., Rab GTPases, SNAREs, annexins, and flotillin), proteins involved in EV biogenesis (e.g., Alix and Tsg101), transmembrane proteins composed of four domains such as tetraspanins (CD63, CD81, CD9 and CD82), major histocompatibility complex (MHC) molecules and cytosolic proteins such as heat shock proteins [20]. EVs also present a lipidic component. Exosomes are enriched in cholesterol, sphingomyelin, phosphatidylcholine, and phosphatidylethanolamine, as well as proteins associated to lipid rafts such as glycosylphosphatidylinositol-anchored proteins and flotillin. Nevertheless, the lipid content of microvesicles is less known than exosomes and is primarily involved in their budding from the plasma membrane [75]. Regarding nucleic acids, it was reported that microvesicles and exosomes carry single-stranded DNA (ssDNA), double-stranded DNA (dsDNA), genomic DNA, mitochondrial DNA, and reverse-transcribed complementary DNAs [76,77]. On the other hand, EVs can deliver mRNA, mRNA fragments, long non-coding RNA (lnRNA), miRNA, ribosomal RNA (rRNA) and fragments of tRNA, vault- and Y-RNA [78]. More importantly, miRNAs may act on the behavior of recipient cells by regulating gene expression, immune response, and additional physiological processes [79,80].

Exosomes and microvesicles have two different pathways of biogenesis. They are constantly produced by viable cells, while apoptotic bodies are released exclusively during the apoptotic phase of the cell cycle [81]. Whereas the releasing mechanisms of apoptotic bodies from the cell membrane have been widely elucidated, those related to microvesicles have been investigated only recently. Microvesicles are formed by fusion with the internal side of the plasma membrane and requires several rearrangements within the membrane itself such as reorganization of lipid and protein components and Ca^{2+} levels [82,83]. It has been reported that a reduction in cholesterol in microvesicles may compromise their generation in activated neutrophils [84]. In addition, the small GTPase family Rho and the protein kinase associated with Rho (ROCK) induce the biogenesis of microvesicles in various populations of cancer cells by regulating the activity of cytoskeletal components such as actin [85].

Exosomes are defined as intraluminal vesicles (ILVs) formed by the invagination of the endosomal membrane during the maturation of multivesicular endosomes (MVEs). This process leads to the acquisition of cargo and emission of vesicles in the lumen through particular sorting machineries (Figure 3) [86].

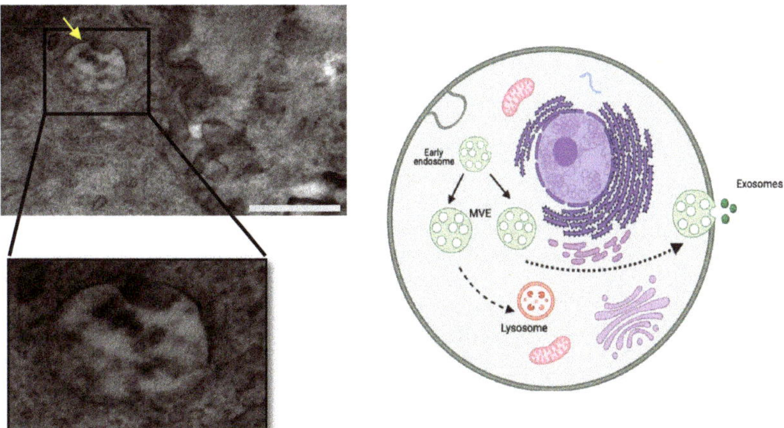

Figure 3. Visualization of MVEs by electron microscopy (scale bar = 1 μm). Biogenesis process of ILVs by budding into early endosomes. MVEs release exosomes by fusion with the plasma membrane or fuse with lysosomes. ILVs = intraluminal vesicles; MVEs = multivesicular endosomes. Created with BioRender.com (accessed on 10 October 2021).

The discovery of the endosomal sorting complex required for transport (ESCRT) machinery was important as it seems to be a fundamental component in several cell types for the generation of MVEs and ILVs [87,88]. The first step is the enrolling of ESCRT-0 subunits, which select cargo in a ubiquitin-dependent manner; ESCRT-I and ESCRT–II are involved in budding endosomal membrane formation, and ESCRT-III is deputed to ILV scission and recycling of the ESCRT machinery [89]. Tumor susceptibility gene 101 (TSG101) and ALIX are two important members of the ESCRT complex discovered by Théry et al. in a proteomic analysis of mouse dendritic cell-derived exosomes [90].

Exosomes can also be formed through an ESCRT-independent mechanism. The sphingolipid ceramide plays a role by creating a spontaneous negative curvature in the membrane leaflets of MVEs to shape ILVs [91]. The formation and the sorting of ILV cargo can also be regulated by the syndecan–syntenin–ALIX pathway. Indeed, syndecan heparan sulphate proteoglycans are connected to their cytoplasmic adaptor syntenin, which interacts with ALIX hence promoting endosomal membrane budding [92].

Finally, exosomal surface proteins belonging to the tetraspanine family, such as CD63 CD81, CD82 and CD9 or heat shock protein (Hsp70), may control ESCRT-independent exosomes biogenesis [93,94]. Intracellular vesicular trafficking, fusion with the plasma membrane, and consequently, exosome secretion are driven by Rab GTPases proteins [95]. Mechanisms regulating the sorting of nucleic acids in exosomes are still not known.

In summary, microvesicles bud directly from the plasma membrane, while exosomes are firstly incorporated into the endosomal vesicles and then released into the extracellular milieu by fusion of MVEs with the plasma membrane. Once released, EVs are absorbed by neighboring cells by endocytosis, phagocytosis, pinocytosis, or fusion of membranes. Ligands on the extracellular vesicle membrane can also bind to a receptor on target cells to thereby induce an intracellular signal [96].

5. Exosome Therapy for IVD Regeneration

Emerging cell therapies have demonstrated promising results by supplementing damaged IVDs with stem cells harvested from different sources [97]. Despite this, there is a growing interest in cell-free therapy. Recent research has proposed that exosomes derived from different tissues such as bone marrow and adipose tissue could represent an emerging tool in regenerative medicine [98]. The use of MSC-derived exosomes may offer several advantages. In fact, they do not require tissue engraftment but present high biocompatibility, low immunogenicity and toxicity, less oncogenic potential and substantial economic benefits compared to cell therapy. Such aspects may promote the use of MSCs-derived exosomes for IVD regeneration as an attractive alternative to intradiscal cell therapy [99].

However, before the exploitation and application of exosomes, it is necessary to isolate them efficiently from biological fluids or supernatants from in vitro cell cultures [100,101]. Ultracentrifugation is the most used technique and is often coupled with ultrafiltration to improve the purity of the sample, although this provides the risk of nanoparticle deformation [102,103]. Other sophisticated approaches include tangential flow filtration, dimensional exclusion chromatography and polymeric precipitation [104–106]. Several methods are available for exosome characterization. Transmission electron microscopy (TEM) and scanning electron microscopy (SEM) are often employed to verify their morphology. Nanoparticle tracking analyzers (NTA) measure the diameter size and specific EV markers (CD9, CD63, CD81, TSG101 and ALIX) can be identified through flow cytometry [107].

The role of exosomes has been evaluated with regards to ECM synthesis, metabolic dysfunction, anti-apoptotic processes, inflammation, and wound healing (Table 2) [108,109].

Table 2. Summary of available evidence on the role of EVs in IDD.

Authors	Type of Study	Potential Role of EVs	Results
Liao et al. [110]	In vitro/In vivo	Modulation of ER stress and reduction in NPC apoptosis during IDD	Attenuation of ER stress-induced apoptosis by activating AKT and ERK signaling in vitro and in a rat tail model
Lan et al. [111]	In vitro/In vivo	Induction of NPC exosomes on the differentiation of MSCs into NP-like cells	NPC exosomes play a key role in the differentiation of MSCs into NP-like cells through inhibition of the Notch1 pathway
Lu et al. [112]	In vitro	Intercellular communication between BM-MSCs and NPCs after exosomal reciprocal up-take	Improvement in BM-MSC migration and differentiation to a NP-like phenotype, NPC proliferation and ECM production
Xia et al. [113]	In vitro/In vivo	Restorative effects of exosomes on H_2O_2-induced NPC inflammation, ROS production and mitochondrial dysfunction	Anti-inflammatory role in pathological NPCs and reversion of the damaged mitochondria. In the rabbit model exosomes significantly prevented IDD
Hingert et al. [114]	In vitro	Stimulation of ECM production, cell proliferation and reduction in cell apoptosis in disc cells after treatment with EVs from hMSCs	hMSCs-derived EVs increased cell viability and stimulated chondrogenesis in NPCs from degenerated IVDs
Li et al. [115]	In vitro	Protective effect on human NPCs in the IVD acid environment	MSC-derived exosomes promote the expression of ECM, protecting NPCs from acidic pH-induced damage and apoptosis
Li et al. [116]	In vitro	BM-MSC-exosomes promoted AF cells proliferation	Suppression of IL-1β-induced inflammation, apoptosis rate and autophagy in AF cells
Luo et al. [117]	In vitro/In vivo	CEP stem cell-derived exosomes decrease the apoptosis rate of NPCs in vitro and activated autophagy in vivo	Reduction in the apoptotic NPCs after treatment with healthy exosomes compared to degenerated exosomes and attenuated IDD via activation of the AKT and autophagy pathways
Luo et al. [118]	In vitro/In vivo	Stimulation of the migration of CEP stem cells into the IVD and their differentiation into NPCs via autocrine exosomes release	CEP stem cell-derived exosomes promoted invasion, migration, and differentiation of cartilage endplate stem cells by autocrine exosomes via the HIF-1α/Wnt pathway
Tang et al. [119]	In vitro/In vivo	Engineered EVs deliver FOXF1 and reprogram human NPCs in vitro and mouse IVD cells in vivo	Upregulation of healthy NP markers (FOXF1, KRT19), downregulation of IL-1β and IL-6, MMP13, and NGF, with significant increases of GAG in human NPCs
Sun et al. [120]	In vitro	AF-derived exosomes mediate intercellular communication between AF cells and HUVECs	AF- derived exosomes could be phagocytosed by HUVECs regulating cell migration and inflammation with a proangiogenic effect in IDD

AF = annulus fibrosus; BM-MSCs = bone marrow-derived mesenchymal stem cells; CEP = cartilaginous endplate; ECM = extracellular matrix; ER = endoplasmic reticulum; EVs = extracellular vesicles; GAG = glycosaminoglycan; hMSCs = human mesenchymal stem cells; HUVECs = human umbilical vein endothelial cells; IDD = intervertebral disc degeneration; IL = interleukin; IVD = intervertebral disc; MMP = matrix metalloproteinase; MSC = mesenchymal stem cell; NGF = nerve growth factor; NP = nucleus pulposus; NPC = nucleopulpocyte; ROS = reactive oxygen species.

The beneficial effect of MSC-derived exosomes on NPCs in acidic conditions has been recently demonstrated. BM-MSC-exosomes increased proliferation and attenuated the apoptosis of NPCs and enhanced the expression of ECM components [115]. Moreover, exosomes are able to inhibit IL-1β-induced inflammation and apoptosis in AF cells and they promote cell proliferation [116]. On the other hand, Liu et al. identified various similarities between CEP stem cells and MSCs, including their capacity to communicate with IVD cells by secreting extracellular vesicles [121]. Indeed, IVD repair may be promoted by CEP stem cell exosomes by inhibiting NPC apoptosis and encouraging the migration of CEP stem cells into the IVD and differentiate into NPCs via autocrine mechanisms [117,118]. AF-derived exosomes have also showed a protective effect in IDD [120]. Aging and degenerative tissues usually accumulate in endoplasmic reticulum (ER) advanced glycation end-products (AGEs) related to inflammation, cell death, metabolic dysfunction, and stress [122,123]. Liao et al. found that the injection of MSCs-derived exosomes in an in vivo rat IDD model modulated ER stress by protecting NPCs against cell death and ameliorating IDD through an attenuation of AGE-induced ER stress mediated by AKT and ERK pathways [110]. Exosomes may also induce differentiation of stem cells towards different phenotypes, although the underlying mechanisms are still unknown. It was reported that the inhibition of Notch1 may promote exosome-induced differentiation of MSCs into NP-like cells in vitro [111]. Lu et al. analyzed exosomes secreted by both NP cells and BM-MSCs in a co-culture system. They highlighted that NPC-derived exosomes induced BM-MSC differentiation towards a NP-like phenotype, whereas BM-MSC-derived exosomes promoted NPC proliferation and ECM production [112].

Recently, a study by Xia et al. described the protective role of exosomes in a rabbit IDD model. Exosomes inhibited H_2O_2-induced NLRP3 inflammasome activation, drastically reduced the catabolic action of MMP-3, MMP-13, IL-6 and inducible nitric oxide synthase (iNOS), repressed reactive oxygen species (ROS) generation and mitochondrial dysfunction was restored [113]. Hingert et al. have reported the potential abilities of MSC-derived EVs in a 3D in vitro model to improve cell viability, proliferation, ECM production, apoptosis, chondrogenesis and cytokine secretion in NPCs harvested from patients with IDD and chronic LBP [114]. Furthermore, an actual challenge is the potential of engineered EVs to lead NPC towards a healthy NP-like phenotype, decrease catabolism and inflammation, and improve GAG accumulation both in vitro and in vivo.

However, the clinical application of exosomes is still premature, and appropriate strategies will be necessary to exploit exosomes as drug delivery vehicles with high specificity, non-cytotoxic effects and low immunogenicity [67].

6. The Role of miRNAs in IDD

Exosomes may be promising diagnostic biomarkers due to their ability to incorporate several molecules including lipids, RNAs and proteins targeting specific recipient cells or tissues [124]. Interestingly, many studies have shown that several miRNAs can be predictive markers for IDDs. miRNAs are single-stranded and small non-coding RNAs that can bind to the 3′-untranslated region (3′-UTR) of target mRNA molecules, leading to their translational repression or degradation [125]. They represent key biomolecular players that regulate the expression of several target genes and are involved through various pathways in cell proliferation, apoptosis, immunomodulation, ECM anabolism and catabolism (Figure 4) [126].

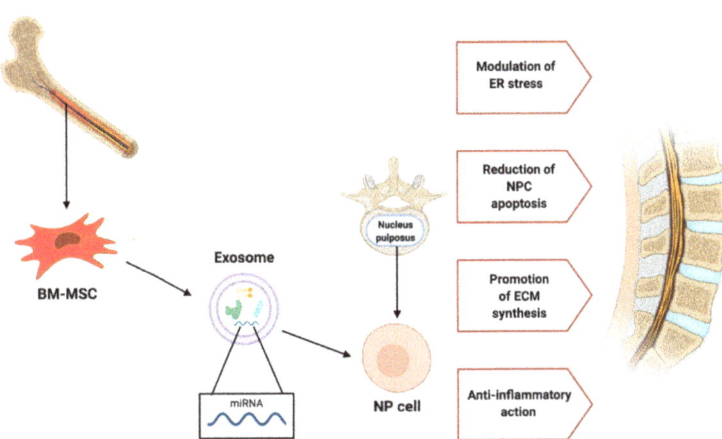

Figure 4. Potential therapeutic role of exosomes in IDD through the modulation of ER stress, stimulation of production of ECM components, anti-apoptotic and anti-inflammatory effects on NP cells. BM-MSCs = bone marrow-derived stem cells; ECM = extracellular matrix; ER = endoplasmic reticulum; NP = nucleus pulposus; NPC = nucleus pulposus cell. Created with BioRender.com (accessed on 10 October 2021).

The anti-apoptotic effects of exosomes from MSCs-derived CM were described in a study conducted by Cheng et al. These results seem to be partially mediated by the transfer of exosomal miR-21 associated with the downregulation of homologous phosphatase and tensin (PTEN), which leads to the activation of the phosphoinositide 3-kinase (PI3K)/Akt pathway and the inhibition of TNF-α-induced NPC apoptosis [127]. The regulatory role of miR-210 was also investigated. Indeed, miR-210 upregulation inhibited apoptosis in human NPCs, suggesting its involvement in the etiology of IDD [128]. It has hypothesized that increased miR-223 expression may be a predictive marker of chronic lumbar radicular pain, as it would be released from the NP following disc herniation [129]. Similarly, the transfection of miR-146a in bovine NPCs decreased IL-1β-driven inflammatory and catabolic responses both in vitro and ex vivo [130]. Chen et al. have reported that miR-494-5p was upregulated in NP tissues of a mouse IDD model, while tissue inhibitors of metalloproteinases-3 (TIMP-3) were downregulated. The reduced expression of this miRNA promoted viability and attenuated the apoptosis and senescence of NPCs [131]. The upregulation of miR-142-3p or miR-199 delivered by MSC-derived exosomes may protect NPC from injury through inhibition of mitogen-activated protein kinase (MAPK) signaling [132,133]. In addition, miR-199a decreased MMP-2, MMP-6, TIMP-1 and apoptotic cells in a mouse model of IDD, while the expression of collagen type II, aggrecan, increased following the exposition to BMSC EVs in vitro [134]. Among other miRNAs previously identified, miR-27a was involved in the regulation of apoptotic pathways and in the prevention of IDD through the ability to suppress ECM degradation targeting MMP-13 [135,136]. Similarly, under TNF-α stimulation, exosomes extracted from BM-MSCs were able to secrete miR-532-5p. This exosomal miRNA has been linked to a reduction in apoptosis in NP cells and matrix degradation by targeting Ras association domain family 5 (RASSF5), attenuating disc degeneration [137]. Zhang et al. suggested the potential effect of MSC-derived exosomes on pyroptosis activation and the role of miR-410 as a specific mediator of NPC pyroptosis [138].

Activating transcription factor 6 (ATF6) is a transcription factor identified as the target gene of miR-31-5p. When miR-31-5p was upregulated, the expression of ATF6 and other proteins involved in oxidative stress-induced ER-stress was suppressed, resulting in a protective effect in IDD [139].

Another family of molecules delivered by exosomes is represented by circular noncoding RNAs (circRNAs). CircRNAs may act as post-transcriptional regulators and interact with miRNAs as miRNA sponges. In a recent study, the role of exosomal circRNA_0000253 was explored, suggesting that it may competitively absorb miRNA-141-5p and downregulate SIRT1 promoting IDD progression [140].

Further research is required to optimize the effects of MSCs-derived exosomes and to confirm the association of their miRNAs with the pathogenesis of IDD. It will be crucial to know the active components responsible for their actions, underlying mechanisms, pharmacokinetic aspects, and safety of this approach. Table 3 summarizes the different miRNA studied in IVD research.

Table 3. Summary of available evidence on the role of miRNA in IDD.

Authors	miRNA	Target Gene	Results
Cheng et al. [127]	miR-21	PTEN	miR-21 alleviated TNF-α induced NPC apoptosis by inhibiting PTEN
Zhang et al. [128]	miR-210	HOXA9	miR-210 significantly decreased NPC apoptosis via downregulation of HOXA9
Moen et al. [129]	miR-223	-	miR-223 may be a predictive biomarker associated with persistent lumbar radicular pain
Gu et al. [130]	miR-146a	TRAF6/IRAK-1	An increase in miR-146a protects against IL-1 induced IVD degeneration and inflammation
Chen et al. [131]	miR-494-5p	TIMP-3	A dysregulation of miR-494-5p could decelerate the progression of IDD via promoting the expression of TIMP3
Zhu et al. [132]	miR-142-3p	MLK3	Inhibition of inflammation, cell apoptosis, and MAPK signaling activation in NPCs through the downregulation of MLK3
Wang et al. [133]	miR-199	MAP3K5	TNF-α induced NPC apoptosis by downregulating MAP3K5
Wen et al. [134]	miR-199a	GREM1	miR-199a carried by BM-MSC EVs promotes IVD repair by targeting GREM1 and downregulating the TGF-β pathway
Liu et al. [135]	miR-27a	PI3K/MMP-13	Upregulation of miR-27a promotes apoptosis in human IVD by targeting PI3K
Zhang et al. [136]	miR-27a	PI3K/MMP-13	miR-27a suppressed ECM degradation and MMP-13 expression induced by IL-1β in NP cells
Zhu et al. [137]	miR-532-5p	RASSF5	Suppression of TNF-α-induced apoptosis, ECM degradation, and fibrosis in NPCs
Zhang et al. [138]	miR-140	NLRP3	miR-410 delivered by MSC exosomes may inhibit NLRP3 inflammasome-mediated pyroptosis in NPCs
Xie et al. [139]	miR-31-5p	ATF6	MSC exosomes negatively regulated ER stress, apoptosis and calcification in CEP through miR-31-5p

ATF = activating transcription factor; BM-MSCs = bone marrow-derived mesenchymal stem cells; CEP = cartilaginous endplate; ECM = extracellular matrix; EVs = extracellular vesicles; GREM = gremlin; HOX = homeobox; IL = interleukin; IRAK = interleukin 1 receptor associated kinase; IDD = intervertebral disc degeneration; IVD = intervertebral disc; MAPK = mitogen-activated protein kinase; MLK = mixed-lineage protein kinase; MMP = matrix metalloproteinase; NLRP = nucleotide-binding oligomerization domain, leucine rich repeat and pyrin domain containing; NPC = nucleus pulposus cell; PI3K = phosphoinositide 3-kinase; PTEN = phosphatase and tensin homolog; RASSF = Ras association domain family; TGF = transforming growth factor; TIMP = tissue inhibitor of metalloproteinases; TNF = tumor necrosis factor; TRAF = tumor necrosis factor receptor-associated factor.

7. Conclusions

In recent years, the transplantation of adult MSCs has been considered a promising approach for IVD tissue regeneration due to their capacity to secrete growth factors, cytokines, and other molecules with paracrine activity. However, the success of MSC-based therapies depends on the survival rate and functionality of transplanted cells within the harsh IVD microenvironment. A growing number of clinical trials have opted for cell therapy, even though there are still many doubts about their routine clinical application.

The use of cell-free therapies could circumvent many obstacles. Numerous studies have recently shown that EVs are released by many cell types including MSCs. These small vesicles can transfer functional proteins, RNA, miRNAs, and lipids among cells, hence promoting intercellular communication, proliferation, inhibition of apoptosis and tissue damage repair. Exosomes represent a promising alternative to traditional stem cell-based therapies. Interestingly, exosomal miRNAs may be predictive biomarkers that regulate the balance between regeneration and degeneration mechanisms in IDD. However, due to many unknown pathophysiological processes in which exosomes are involved, further studies are necessary to confirm the ability of these nanoparticles to promote IVD health and homeostasis.

Author Contributions: Conceptualization, V.T., G.V.; methodology, V.T.; data curation, V.T., C.C., G.D.G.; writing—original draft preparation, V.T.; writing—review and editing, G.V., L.A., F.R., C.C., G.D.G.; supervision, R.P., V.D. All authors have read and agreed to the published version of the manuscript.

Funding: This research received no external funding.

Institutional Review Board Statement: Not applicable.

Informed Consent Statement: Not applicable.

Data Availability Statement: The datasets used and/or analyzed during the current study are available from the corresponding author on reasonable request.

Acknowledgments: The support by the ACTIVE (ClinicalTrials.gov Identifier: NCT04759105), DREAM (ClinicalTrials.gov Identifier: NCT05066334), RESPINE (ClinicalTrials.gov Identifier: NCT03737461) and iPSspine (European Union's Horizon 2020 Research and Innovation Programme under grant agreement no. 825925) projects are gratefully acknowledged.

Conflicts of Interest: The authors declare no conflict of interest.

References

1. Hoy, D.; Bain, C.; Williams, G.; March, L.; Brooks, P.; Blyth, F.; Woolf, A.; Vos, T.; Buchbinder, R. A systematic review of the global prevalence of low back pain. *Arthritis Rheum* **2012**, *64*, 2028–2037. [CrossRef]
2. Fernandez-Moure, J.; Moore, C.A.; Kim, K.; Karim, A.; Smith, K.; Barbosa, Z.; Van Eps, J.; Rameshwar, P.; Weiner, B. Novel therapeutic strategies for degenerative disc disease: Review of cell biology and intervertebral disc cell therapy. *SAGE Open Med.* **2018**, *6*, 2050312118761674. [CrossRef] [PubMed]
3. Schneider, B.J.; Haring, R.S.; Song, A.; Kim, P.; Ayers, G.D.; Kennedy, D.J.; Jain, N.B. Characteristics of ambulatory spine care visits in the United States, 2009–2016. *J. Back Musculoskelet. Rehabil.* **2021**, *34*, 657–664. [CrossRef]
4. Russo, F.; De Salvatore, S.; Ambrosio, L.; Vadala, G.; Fontana, L.; Papalia, R.; Rantanen, J.; Iavicoli, S.; Denaro, V. Does Workers' Compensation Status Affect Outcomes after Lumbar Spine Surgery? A Systematic Review and Meta-Analysis. *Int. J. Environ. Res. Public Health* **2021**, *18*, 6165. [CrossRef]
5. Cassidy, J.J.; Hiltner, A.; Baer, E. Hierarchical structure of the intervertebral disc. *Connect. Tissue Res.* **1989**, *23*, 75–88. [CrossRef] [PubMed]
6. Bruehlmann, S.B.; Rattner, J.B.; Matyas, J.R.; Duncan, N.A. Regional variations in the cellular matrix of the annulus fibrosus of the intervertebral disc. *J. Anat.* **2002**, *201*, 159–171. [CrossRef] [PubMed]
7. Trout, J.J.; Buckwalter, J.A.; Moore, K.C.; Landas, S.K. Ultrastructure of the human intervertebral disc. I. Changes in notochordal cells with age. *Tissue Cell* **1982**, *14*, 359–369. [CrossRef]
8. Risbud, M.V.; Guttapalli, A.; Tsai, T.T.; Lee, J.Y.; Danielson, K.G.; Vaccaro, A.R.; Albert, T.J.; Gazit, Z.; Gazit, D.; Shapiro, I.M. Evidence for skeletal progenitor cells in the degenerate human intervertebral disc. *Spine (Phila Pa 1976)* **2007**, *32*, 2537–2544. [CrossRef]

9. Vadala, G.; Russo, F.; Ambrosio, L.; Loppini, M.; Denaro, V. Stem cells sources for intervertebral disc regeneration. *World J. Stem Cells* **2016**, *8*, 185–201. [CrossRef]
10. Vadala, G.; Russo, F.; Di Martino, A.; Denaro, V. Intervertebral disc regeneration: From the degenerative cascade to molecular therapy and tissue engineering. *J. Tissue Eng. Regen Med.* **2015**, *9*, 679–690. [CrossRef]
11. Sowa, G.; Westrick, E.; Pacek, C.; Coelho, P.; Patel, D.; Vadala, G.; Georgescu, H.; Vo, N.; Studer, R.; Kang, J. In vitro and in vivo testing of a novel regulatory system for gene therapy for intervertebral disc degeneration. *Spine (Phila Pa 1976)* **2011**, *36*, E623–E628. [CrossRef]
12. Hubert, M.G.; Vadala, G.; Sowa, G.; Studer, R.K.; Kang, J.D. Gene therapy for the treatment of degenerative disk disease. *J. Am. Acad. Orthop. Surg.* **2008**, *16*, 312–319. [CrossRef]
13. Vadala, G.; Russo, F.; Musumeci, M.; D'Este, M.; Cattani, C.; Catanzaro, G.; Tirindelli, M.C.; Lazzari, L.; Alini, M.; Giordano, R.; et al. Clinically relevant hydrogel-based on hyaluronic acid and platelet rich plasma as a carrier for mesenchymal stem cells: Rheological and biological characterization. *J. Orthop. Res.* **2017**, *35*, 2109–2116. [CrossRef] [PubMed]
14. Russo, F.; Ambrosio, L.; Peroglio, M.; Guo, W.; Wangler, S.; Gewiess, J.; Grad, S.; Alini, M.; Papalia, R.; Vadala, G.; et al. A Hyaluronan and Platelet-Rich Plasma Hydrogel for Mesenchymal Stem Cell Delivery in the Intervertebral Disc: An Organ Culture Study. *Int. J. Mol. Sci.* **2021**, *22*, 2963. [CrossRef] [PubMed]
15. Vadala, G.; Sowa, G.A.; Smith, L.; Hubert, M.G.; Levicoff, E.A.; Denaro, V.; Gilbertson, L.G.; Kang, J.D. Regulation of transgene expression using an inducible system for improved safety of intervertebral disc gene therapy. *Spine (Phila Pa 1976)* **2007**, *32*, 1381–1387. [CrossRef]
16. Vadala, G.; Sowa, G.A.; Kang, J.D. Gene therapy for disc degeneration. *Expert Opin. Biol. Ther.* **2007**, *7*, 185–196. [CrossRef] [PubMed]
17. Vadala, G.; Ambrosio, L.; Russo, F.; Papalia, R.; Denaro, V. Interaction between Mesenchymal Stem Cells and Intervertebral Disc Microenvironment: From Cell Therapy to Tissue Engineering. *Stem Cells Int.* **2019**, *2019*, 2376172. [CrossRef]
18. Cai, Q.; Zhu, A.; Gong, L. Exosomes of glioma cells deliver miR-148a to promote proliferation and metastasis of glioblastoma via targeting CADM1. *Bull. Cancer* **2018**, *105*, 643–651. [CrossRef]
19. Kim, K.H.; Jo, J.H.; Cho, H.J.; Park, T.S.; Kim, T.M. Therapeutic potential of stem cell-derived extracellular vesicles in osteoarthritis: Preclinical study findings. *Lab. Anim Res.* **2020**, *36*, 10. [CrossRef]
20. Yanez-Mo, M.; Siljander, P.R.; Andreu, Z.; Zavec, A.B.; Borras, F.E.; Buzas, E.I.; Buzas, K.; Casal, E.; Cappello, F.; Carvalho, J.; et al. Biological properties of extracellular vesicles and their physiological functions. *J. Extracell Vesicles* **2015**, *4*, 27066. [CrossRef]
21. Qin, Y.; Sun, R.; Wu, C.; Wang, L.; Zhang, C. Exosome: A Novel Approach to Stimulate Bone Regeneration through Regulation of Osteogenesis and Angiogenesis. *Int. J. Mol. Sci.* **2016**, *17*, 712. [CrossRef] [PubMed]
22. Wu, X.; Wang, Y.; Xiao, Y.; Crawford, R.; Mao, X.; Prasadam, I. Extracellular vesicles: Potential role in osteoarthritis regenerative medicine. *J. Orthop. Transl.* **2020**, *21*, 73–80. [CrossRef]
23. Pattappa, G.; Li, Z.; Peroglio, M.; Wismer, N.; Alini, M.; Grad, S. Diversity of intervertebral disc cells: Phenotype and function. *J. Anat.* **2012**, *221*, 480–496. [CrossRef] [PubMed]
24. Rajasekaran, S.; Vidyadhara, S.; Subbiah, M.; Kamath, V.; Karunanithi, R.; Shetty, A.P.; Venkateswaran, K.; Babu, M.; Meenakshi, J. ISSLS prize winner: A study of effects of in vivo mechanical forces on human lumbar discs with scoliotic disc as a biological model: Results from serial postcontrast diffusion studies, histopathology and biochemical analysis of twenty-one human lumbar scoliotic discs. *Spine (Phila Pa 1976)* **2010**, *35*, 1930–1943. [CrossRef]
25. Bartels, E.M.; Fairbank, J.C.; Winlove, C.P.; Urban, J.P. Oxygen and lactate concentrations measured in vivo in the intervertebral discs of patients with scoliosis and back pain. *Spine (Phila Pa 1976)* **1998**, *23*, 1–7. [CrossRef]
26. Selard, E.; Shirazi-Adl, A.; Urban, J.P. Finite element study of nutrient diffusion in the human intervertebral disc. *Spine (Phila Pa 1976)* **2003**, *28*, 1945–1953. [CrossRef]
27. Bez, M.; Zhou, Z.; Sheyn, D.; Tawackoli, W.; Giaconi, J.C.; Shapiro, G.; Ben David, S.; Gazit, Z.; Pelled, G.; Li, D.; et al. Molecular pain markers correlate with pH-sensitive MRI signal in a pig model of disc degeneration. *Sci. Rep.* **2018**, *8*, 17363. [CrossRef]
28. Molinos, M.; Almeida, C.R.; Caldeira, J.; Cunha, C.; Goncalves, R.M.; Barbosa, M.A. Inflammation in intervertebral disc degeneration and regeneration. *J. R. Soc. Interface* **2015**, *12*, 20141191. [CrossRef] [PubMed]
29. Johnson, Z.I.; Shapiro, I.M.; Risbud, M.V. Extracellular osmolarity regulates matrix homeostasis in the intervertebral disc and articular cartilage: Evolving role of TonEBP. *Matrix Biol.* **2014**, *40*, 10–16. [CrossRef]
30. Urban, J.P. The role of the physicochemical environment in determining disc cell behaviour. *Biochem. Soc. Trans.* **2002**, *30*, 858–864. [CrossRef] [PubMed]
31. Wuertz, K.; Urban, J.P.; Klasen, J.; Ignatius, A.; Wilke, H.J.; Claes, L.; Neidlinger-Wilke, C. Influence of extracellular osmolarity and mechanical stimulation on gene expression of intervertebral disc cells. *J. Orthop. Res.* **2007**, *25*, 1513–1522. [CrossRef] [PubMed]
32. Tao, Y.Q.; Liang, C.Z.; Li, H.; Zhang, Y.J.; Li, F.C.; Chen, G.; Chen, Q.X. Potential of co-culture of nucleus pulposus mesenchymal stem cells and nucleus pulposus cells in hyperosmotic microenvironment for intervertebral disc regeneration. *Cell Biol. Int.* **2013**, *37*, 826–834. [CrossRef] [PubMed]
33. Desmoulin, G.T.; Pradhan, V.; Milner, T.E. Mechanical Aspects of Intervertebral Disc Injury and Implications on Biomechanics. *Spine (Phila Pa 1976)* **2020**, *45*, E457–E464. [CrossRef] [PubMed]
34. Fearing, B.V.; Hernandez, P.A.; Setton, L.A.; Chahine, N.O. Mechanotransduction and cell biomechanics of the intervertebral disc. *JOR Spine* **2018**, *1*. [CrossRef] [PubMed]

35. Vo, N.V.; Hartman, R.A.; Patil, P.R.; Risbud, M.V.; Kletsas, D.; Iatridis, J.C.; Hoyland, J.A.; Le Maitre, C.L.; Sowa, G.A.; Kang, J.D. Molecular mechanisms of biological aging in intervertebral discs. *J. Orthop. Res.* **2016**, *34*, 1289–1306. [CrossRef]
36. Risbud, M.V.; Shapiro, I.M. Role of cytokines in intervertebral disc degeneration: Pain and disc content. *Nat. Rev. Rheumatol.* **2014**, *10*, 44–56. [CrossRef]
37. Wang, Y.; Che, M.; Xin, J.; Zheng, Z.; Li, J.; Zhang, S. The role of IL-1beta and TNF-alpha in intervertebral disc degeneration. *Biomed. Pharmacother.* **2020**, *131*, 110660. [CrossRef]
38. Wang, C.; Yu, X.; Yan, Y.; Yang, W.; Zhang, S.; Xiang, Y.; Zhang, J.; Wang, W. Tumor necrosis factor-alpha: A key contributor to intervertebral disc degeneration. *Acta Biochim. Biophys. Sin. (Shanghai)* **2017**, *49*, 1–13. [CrossRef] [PubMed]
39. Kadow, T.; Sowa, G.; Vo, N.; Kang, J.D. Molecular basis of intervertebral disc degeneration and herniations: What are the important translational questions? *Clin. Orthop. Relat. Res.* **2015**, *473*, 1903–1912. [CrossRef]
40. Urban, J.P.; Smith, S.; Fairbank, J.C. Nutrition of the intervertebral disc. *Spine (Phila Pa 1976)* **2004**, *29*, 2700–2709. [CrossRef]
41. Sakai, D.; Nakamura, Y.; Nakai, T.; Mishima, T.; Kato, S.; Grad, S.; Alini, M.; Risbud, M.V.; Chan, D.; Cheah, K.S.; et al. Exhaustion of nucleus pulposus progenitor cells with ageing and degeneration of the intervertebral disc. *Nat. Commun.* **2012**, *3*, 1264. [CrossRef]
42. Acaroglu, E.R.; Iatridis, J.C.; Setton, L.A.; Foster, R.J.; Mow, V.C.; Weidenbaum, M. Degeneration and aging affect the tensile behavior of human lumbar anulus fibrosus. *Spine (Phila Pa 1976)* **1995**, *20*, 2690–2701. [CrossRef]
43. Liu, Y.; Li, Y.; Nan, L.P.; Wang, F.; Zhou, S.F.; Feng, X.M.; Liu, H.; Zhang, L. Insights of stem cell-based endogenous repair of intervertebral disc degeneration. *World J. Stem Cells* **2020**, *12*, 266–276. [CrossRef]
44. Sobajima, S.; Vadala, G.; Shimer, A.; Kim, J.S.; Gilbertson, L.G.; Kang, J.D. Feasibility of a stem cell therapy for intervertebral disc degeneration. *Spine J.* **2008**, *8*, 888–896. [CrossRef] [PubMed]
45. Cunha, C.; Almeida, C.R.; Almeida, M.I.; Silva, A.M.; Molinos, M.; Lamas, S.; Pereira, C.L.; Teixeira, G.Q.; Monteiro, A.T.; Santos, S.G.; et al. Systemic Delivery of Bone Marrow Mesenchymal Stem Cells for In Situ Intervertebral Disc Regeneration. *Stem Cells Transl. Med.* **2017**, *6*, 1029–1039. [CrossRef] [PubMed]
46. Wangler, S.; Peroglio, M.; Menzel, U.; Benneker, L.M.; Haglund, L.; Sakai, D.; Alini, M.; Grad, S. Mesenchymal Stem Cell Homing Into Intervertebral Discs Enhances the Tie2-positive Progenitor Cell Population, Prevents Cell Death, and Induces a Proliferative Response. *Spine (Phila Pa 1976)* **2019**, *44*, 1613–1622. [CrossRef] [PubMed]
47. Vadala, G.; Studer, R.K.; Sowa, G.; Spiezia, F.; Iucu, C.; Denaro, V.; Gilbertson, L.G.; Kang, J.D. Coculture of bone marrow mesenchymal stem cells and nucleus pulposus cells modulate gene expression profile without cell fusion. *Spine (Phila Pa 1976)* **2008**, *33*, 870–876. [CrossRef]
48. Vadala, G.; Sobajima, S.; Lee, J.Y.; Huard, J.; Denaro, V.; Kang, J.D.; Gilbertson, L.G. In vitro interaction between muscle-derived stem cells and nucleus pulposus cells. *Spine J.* **2008**, *8*, 804–809. [CrossRef]
49. Vadala, G.; Ambrosio, L.; Russo, F.; Papalia, R.; Denaro, V. Stem Cells and Intervertebral Disc Regeneration Overview-What They Can and Can't Do. *Int. J. Spine Surg.* **2021**, *15*, 40–53. [CrossRef]
50. Orozco, L.; Soler, R.; Morera, C.; Alberca, M.; Sanchez, A.; Garcia-Sancho, J. Intervertebral disc repair by autologous mesenchymal bone marrow cells: A pilot study. *Transplantation* **2011**, *92*, 822–828. [CrossRef]
51. Noriega, D.C.; Ardura, F.; Hernandez-Ramajo, R.; Martin-Ferrero, M.A.; Sanchez-Lite, I.; Toribio, B.; Alberca, M.; Garcia, V.; Moraleda, J.M.; Sanchez, A.; et al. Intervertebral Disc Repair by Allogeneic Mesenchymal Bone Marrow Cells: A Randomized Controlled Trial. *Transplantation* **2017**, *101*, 1945–1951. [CrossRef]
52. Rani, S.; Ryan, A.E.; Griffin, M.D.; Ritter, T. Mesenchymal Stem Cell-derived Extracellular Vesicles: Toward Cell-free Therapeutic Applications. *Mol. Ther.* **2015**, *23*, 812–823. [CrossRef]
53. Gnecchi, M.; Danieli, P.; Malpasso, G.; Ciuffreda, M.C. Paracrine Mechanisms of Mesenchymal Stem Cells in Tissue Repair. *Methods Mol. Biol.* **2016**, *1416*, 123–146. [CrossRef] [PubMed]
54. Harrell, C.R.; Fellabaum, C.; Jovicic, N.; Djonov, V.; Arsenijevic, N.; Volarevic, V. Molecular Mechanisms Responsible for Therapeutic Potential of Mesenchymal Stem Cell-Derived Secretome. *Cells* **2019**, *8*, 467. [CrossRef]
55. Wu, L.; Leijten, J.C.; Georgi, N.; Post, J.N.; van Blitterswijk, C.A.; Karperien, M. Trophic effects of mesenchymal stem cells increase chondrocyte proliferation and matrix formation. *Tissue Eng. Part A* **2011**, *17*, 1425–1436. [CrossRef] [PubMed]
56. Teixeira, G.Q.; Pereira, C.L.; Ferreira, J.R.; Maia, A.F.; Gomez-Lazaro, M.; Barbosa, M.A.; Neidlinger-Wilke, C.; Goncalves, R.M. Immunomodulation of Human Mesenchymal Stem/Stromal Cells in Intervertebral Disc Degeneration: Insights From a Proinflammatory/Degenerative Ex Vivo Model. *Spine (Phila Pa 1976)* **2018**, *43*, E673–E682. [CrossRef]
57. Eseonu, O.I.; De Bari, C. Homing of mesenchymal stem cells: Mechanistic or stochastic? Implications for targeted delivery in arthritis. *Rheumatology* **2015**, *54*, 210–218. [CrossRef]
58. Sakai, D.; Schol, J. Cell therapy for intervertebral disc repair: Clinical perspective. *J. Orthop. Transl.* **2017**, *9*, 8–18. [CrossRef]
59. Shim, E.K.; Lee, J.S.; Kim, D.E.; Kim, S.K.; Jung, B.J.; Choi, E.Y.; Kim, C.S. Autogenous Mesenchymal Stem Cells from the Vertebral Body Enhance Intervertebral Disc Regeneration via Paracrine Interaction: An in Vitro Pilot Study. *Cell Transplant.* **2016**, *25*, 1819–1832. [CrossRef] [PubMed]
60. Li, X.; Wu, A.; Han, C.; Chen, C.; Zhou, T.; Zhang, K.; Yang, X.; Chen, Z.; Qin, A.; Tian, H.; et al. Bone marrow-derived mesenchymal stem cells in three-dimensional co-culture attenuate degeneration of nucleus pulposus cells. *Aging (Albany NY)* **2019**, *11*, 9167–9187. [CrossRef]

61. Guerrero, J.; Hackel, S.; Croft, A.S.; Hoppe, S.; Albers, C.E.; Gantenbein, B. The nucleus pulposus microenvironment in the intervertebral disc: The fountain of youth? *Eur. Cell Mater.* **2021**, *41*, 707–738. [CrossRef] [PubMed]
62. Miguelez-Rivera, L.; Perez-Castrillo, S.; Gonzalez-Fernandez, M.L.; Prieto-Fernandez, J.G.; Lopez-Gonzalez, M.E.; Garcia-Cosamalon, J.; Villar-Suarez, V. Immunomodulation of mesenchymal stem cells in discogenic pain. *Spine J.* **2018**, *18*, 330–342. [CrossRef] [PubMed]
63. Loibl, M.; Wuertz-Kozak, K.; Vadalà, G.; Lang, S.; Fairbank, J.; Urban, J.P. Controversies in regenerative medicine: Should intervertebral disc degeneration be treated with mesenchymal stem cells? *JOR Spine* **2019**, *2*, e1043. [CrossRef]
64. Johnstone, R.M.; Adam, M.; Hammond, J.R.; Orr, L.; Turbide, C. Vesicle formation during reticulocyte maturation. Association of plasma membrane activities with released vesicles (exosomes). *J. Biol. Chem.* **1987**, *262*, 9412–9420. [CrossRef]
65. Thery, C.; Witwer, K.W.; Aikawa, E.; Alcaraz, M.J.; Anderson, J.D.; Andriantsitohaina, R.; Antoniou, A.; Arab, T.; Archer, F.; Atkin-Smith, G.K.; et al. Minimal information for studies of extracellular vesicles 2018 (MISEV2018): A position statement of the International Society for Extracellular Vesicles and update of the MISEV2014 guidelines. *J. Extracell Vesicles* **2018**, *7*, 1535750. [CrossRef]
66. Crescitelli, R.; Lasser, C.; Szabo, T.G.; Kittel, A.; Eldh, M.; Dianzani, I.; Buzas, E.I.; Lotvall, J. Distinct RNA profiles in subpopulations of extracellular vesicles: Apoptotic bodies, microvesicles and exosomes. *J. Extracell Vesicles* **2013**, *2*. [CrossRef]
67. Buschmann, D.; Mussack, V.; Byrd, J.B. Separation, characterization, and standardization of extracellular vesicles for drug delivery applications. *Adv. Drug. Deliv. Rev.* **2021**, *174*, 348–368. [CrossRef]
68. Wolf, P. The nature and significance of platelet products in human plasma. *Br. J. Haematol.* **1967**, *13*, 269–288. [CrossRef]
69. Akers, J.C.; Gonda, D.; Kim, R.; Carter, B.S.; Chen, C.C. Biogenesis of extracellular vesicles (EV): Exosomes, microvesicles, retrovirus-like vesicles, and apoptotic bodies. *J. Neurooncol.* **2013**, *113*, 1–11. [CrossRef]
70. Thery, C.; Zitvogel, L.; Amigorena, S. Exosomes: Composition, biogenesis and function. *Nat. Rev. Immunol.* **2002**, *2*, 569–579. [CrossRef]
71. Zhang, Y.; Liu, Y.; Liu, H.; Tang, W.H. Exosomes: Biogenesis, biologic function and clinical potential. *Cell Biosci.* **2019**, *9*, 19. [CrossRef] [PubMed]
72. Zhang, B.; Tian, X.; Hao, J.; Xu, G.; Zhang, W. Mesenchymal Stem Cell-Derived Extracellular Vesicles in Tissue Regeneration. *Cell Transplant.* **2020**, *29*, 963689720908500. [CrossRef]
73. Xu, R.; Greening, D.W.; Zhu, H.J.; Takahashi, N.; Simpson, R.J. Extracellular vesicle isolation and characterization: Toward clinical application. *J. Clin. Investig.* **2016**, *126*, 1152–1162. [CrossRef]
74. Jurj, A.; Zanoaga, O.; Braicu, C.; Lazar, V.; Tomuleasa, C.; Irimie, A.; Berindan-Neagoe, I. A Comprehensive Picture of Extracellular Vesicles and Their Contents. Molecular Transfer to Cancer Cells. *Cancers* **2020**, *12*, 298. [CrossRef]
75. Brouwers, J.F.; Aalberts, M.; Jansen, J.W.; van Niel, G.; Wauben, M.H.; Stout, T.A.; Helms, J.B.; Stoorvogel, W. Distinct lipid compositions of two types of human prostasomes. *Proteomics* **2013**, *13*, 1660–1666. [CrossRef] [PubMed]
76. Balaj, L.; Lessard, R.; Dai, L.; Cho, Y.J.; Pomeroy, S.L.; Breakefield, X.O.; Skog, J. Tumour microvesicles contain retrotransposon elements and amplified oncogene sequences. *Nat. Commun.* **2011**, *2*, 180. [CrossRef] [PubMed]
77. Thakur, B.K.; Zhang, H.; Becker, A.; Matei, I.; Huang, Y.; Costa-Silva, B.; Zheng, Y.; Hoshino, A.; Brazier, H.; Xiang, J.; et al. Double-stranded DNA in exosomes: A novel biomarker in cancer detection. *Cell Res.* **2014**, *24*, 766–769. [CrossRef] [PubMed]
78. Pegtel, D.M.; Gould, S.J. Exosomes. *Annu. Rev. Biochem.* **2019**, *88*, 487–514. [CrossRef]
79. Valadi, H.; Ekstrom, K.; Bossios, A.; Sjostrand, M.; Lee, J.J.; Lotvall, J.O. Exosome-mediated transfer of mRNAs and microRNAs is a novel mechanism of genetic exchange between cells. *Nat. Cell Biol.* **2007**, *9*, 654–659. [CrossRef]
80. Lai, R.C.; Tan, S.S.; Yeo, R.W.; Choo, A.B.; Reiner, A.T.; Su, Y.; Shen, Y.; Fu, Z.; Alexander, L.; Sze, S.K.; et al. MSC secretes at least 3 EV types each with a unique permutation of membrane lipid, protein and RNA. *J. Extracell Vesicles* **2016**, *5*, 29828. [CrossRef]
81. Colombo, M.; Raposo, G.; Thery, C. Biogenesis, secretion, and intercellular interactions of exosomes and other extracellular vesicles. *Annu. Rev. Cell Dev. Biol.* **2014**, *30*, 255–289. [CrossRef] [PubMed]
82. Raposo, G.; Stoorvogel, W. Extracellular vesicles: Exosomes, microvesicles, and friends. *J. Cell Biol.* **2013**, *200*, 373–383. [CrossRef] [PubMed]
83. Piccin, A.; Murphy, W.G.; Smith, O.P. Circulating microparticles: Pathophysiology and clinical implications. *Blood Rev.* **2007**, *21*, 157–171. [CrossRef] [PubMed]
84. del Conde, I.; Shrimpton, C.N.; Thiagarajan, P.; López, J.A. Tissue-factor–bearing microvesicles arise from lipid rafts and fuse with activated platelets to initiate coagulation. *Blood* **2005**, *106*, 1604–1611. [CrossRef] [PubMed]
85. Li, B.; Antonyak, M.A.; Zhang, J.; Cerione, R.A. RhoA triggers a specific signaling pathway that generates transforming microvesicles in cancer cells. *Oncogene* **2012**, *31*, 4740–4749. [CrossRef] [PubMed]
86. Colombo, M.; Moita, C.; van Niel, G.; Kowal, J.; Vigneron, J.; Benaroch, P.; Manel, N.; Moita, L.F.; Thery, C.; Raposo, G. Analysis of ESCRT functions in exosome biogenesis, composition and secretion highlights the heterogeneity of extracellular vesicles. *J. Cell Sci.* **2013**, *126*, 5553–5565. [CrossRef]
87. Tamai, K.; Tanaka, N.; Nakano, T.; Kakazu, E.; Kondo, Y.; Inoue, J.; Shiina, M.; Fukushima, K.; Hoshino, T.; Sano, K.; et al. Exosome secretion of dendritic cells is regulated by Hrs, an ESCRT-0 protein. *Biochem. Biophys. Res. Commun.* **2010**, *399*, 384–390. [CrossRef]
88. Lo Cicero, A.; Stahl, P.D.; Raposo, G. Extracellular vesicles shuffling intercellular messages: For good or for bad. *Curr. Opin. Cell Biol.* **2015**, *35*, 69–77. [CrossRef]

89. Kowal, J.; Tkach, M.; Thery, C. Biogenesis and secretion of exosomes. *Curr. Opin. Cell Biol.* **2014**, *29*, 116–125. [CrossRef]
90. Thery, C.; Boussac, M.; Veron, P.; Ricciardi-Castagnoli, P.; Raposo, G.; Garin, J.; Amigorena, S. Proteomic analysis of dendritic cell-derived exosomes: A secreted subcellular compartment distinct from apoptotic vesicles. *J. Immunol.* **2001**, *166*, 7309–7318. [CrossRef]
91. Trajkovic, K.; Hsu, C.; Chiantia, S.; Rajendran, L.; Wenzel, D.; Wieland, F.; Schwille, P.; Brugger, B.; Simons, M. Ceramide triggers budding of exosome vesicles into multivesicular endosomes. *Science* **2008**, *319*, 1244–1247. [CrossRef]
92. Baietti, M.F.; Zhang, Z.; Mortier, E.; Melchior, A.; Degeest, G.; Geeraerts, A.; Ivarsson, Y.; Depoortere, F.; Coomans, C.; Vermeiren, E.; et al. Syndecan-syntenin-ALIX regulates the biogenesis of exosomes. *Nat. Cell Biol.* **2012**, *14*, 677–685. [CrossRef] [PubMed]
93. Geminard, C.; De Gassart, A.; Blanc, L.; Vidal, M. Degradation of AP2 during reticulocyte maturation enhances binding of hsc70 and Alix to a common site on TFR for sorting into exosomes. *Traffic* **2004**, *5*, 181–193. [CrossRef]
94. van Niel, G.; Charrin, S.; Simoes, S.; Romao, M.; Rochin, L.; Saftig, P.; Marks, M.S.; Rubinstein, E.; Raposo, G. The Tetraspanin CD63 Regulates ESCRT-Independent and -Dependent Endosomal Sorting during Melanogenesis. *Dev. Cell* **2011**, *21*, 708–721. [CrossRef] [PubMed]
95. Ostrowski, M.; Carmo, N.B.; Krumeich, S.; Fanget, I.; Raposo, G.; Savina, A.; Moita, C.F.; Schauer, K.; Hume, A.N.; Freitas, R.P.; et al. Rab27a and Rab27b control different steps of the exosome secretion pathway. *Nat. Cell Biol.* **2010**, *12*, 19–30. [CrossRef]
96. van Niel, G.; D'Angelo, G.; Raposo, G. Shedding light on the cell biology of extracellular vesicles. *Nat. Rev. Mol. Cell Biol.* **2018**, *19*, 213–228. [CrossRef] [PubMed]
97. Alvarez-Viejo, M. Mesenchymal stem cells from different sources and their derived exosomes: A pre-clinical perspective. *World J. Stem Cells* **2020**, *12*, 100–109. [CrossRef]
98. Richardson, S.M.; Kalamegam, G.; Pushparaj, P.N.; Matta, C.; Memic, A.; Khademhosseini, A.; Mobasheri, R.; Poletti, F.L.; Hoyland, J.A.; Mobasheri, A. Mesenchymal stem cells in regenerative medicine: Focus on articular cartilage and intervertebral disc regeneration. *Methods* **2016**, *99*, 69–80. [CrossRef]
99. Elahi, F.M.; Farwell, D.G.; Nolta, J.A.; Anderson, J.D. Preclinical translation of exosomes derived from mesenchymal stem/stromal cells. *Stem Cells* **2020**, *38*, 15–21. [CrossRef]
100. Piazza, N.; Dehghani, M.; Gaborski, T.R.; Wuertz-Kozak, K. Therapeutic Potential of Extracellular Vesicles in Degenerative Diseases of the Intervertebral Disc. *Front. Bioeng. Biotechnol.* **2020**, *8*, 311. [CrossRef]
101. Pinheiro, A.; Silva, A.M.; Teixeira, J.H.; Goncalves, R.M.; Almeida, M.I.; Barbosa, M.A.; Santos, S.G. Extracellular vesicles: Intelligent delivery strategies for therapeutic applications. *J. Control. Release* **2018**, *289*, 56–69. [CrossRef]
102. Thery, C.; Amigorena, S.; Raposo, G.; Clayton, A. Isolation and characterization of exosomes from cell culture supernatants and biological fluids. *Curr. Protoc. Cell Biol.* **2006**, *30*, 3–22. [CrossRef] [PubMed]
103. Witwer, K.W.; Buzas, E.I.; Bemis, L.T.; Bora, A.; Lasser, C.; Lotvall, J.; Nolte-'t Hoen, E.N.; Piper, M.G.; Sivaraman, S.; Skog, J.; et al. Standardization of sample collection, isolation and analysis methods in extracellular vesicle research. *J. Extracell Vesicles* **2013**, *2*. [CrossRef] [PubMed]
104. Corso, G.; Mager, I.; Lee, Y.; Gorgens, A.; Bultema, J.; Giebel, B.; Wood, M.J.A.; Nordin, J.Z.; Andaloussi, S.E. Reproducible and scalable purification of extracellular vesicles using combined bind-elute and size exclusion chromatography. *Sci. Rep.* **2017**, *7*, 11561. [CrossRef]
105. Dehghani, M.; Lucas, K.; Flax, J.; McGrath, J.; Gaborski, T. Tangential flow microfluidics for the capture and release of nanoparticles and extracellular vesicles on conventional and ultrathin membranes. *Adv. Mater. Technol.* **2019**, *4*, 1900539. [CrossRef] [PubMed]
106. Taylor, D.D.; Zacharias, W.; Gercel-Taylor, C. Exosome isolation for proteomic analyses and RNA profiling. *Methods Mol. Biol.* **2011**, *728*, 235–246. [CrossRef] [PubMed]
107. Doyle, L.M.; Wang, M.Z. Overview of Extracellular Vesicles, Their Origin, Composition, Purpose, and Methods for Exosome Isolation and Analysis. *Cells* **2019**, *8*, 727. [CrossRef] [PubMed]
108. Alcaraz, M.J.; Compan, A.; Guillen, M.I. Extracellular Vesicles from Mesenchymal Stem Cells as Novel Treatments for Musculoskeletal Diseases. *Cells* **2019**, *9*, 98. [CrossRef]
109. Shi, Y.; Kang, X.; Wang, Y.; Bian, X.; He, G.; Zhou, M.; Tang, K. Exosomes Derived from Bone Marrow Stromal Cells (BMSCs) Enhance Tendon-Bone Healing by Regulating Macrophage Polarization. *Med. Sci. Monit.* **2020**, *26*, e923328. [CrossRef]
110. Liao, Z.; Luo, R.; Li, G.; Song, Y.; Zhan, S.; Zhao, K.; Hua, W.; Zhang, Y.; Wu, X.; Yang, C. Exosomes from mesenchymal stem cells modulate endoplasmic reticulum stress to protect against nucleus pulposus cell death and ameliorate intervertebral disc degeneration in vivo. *Theranostics* **2019**, *9*, 4084–4100. [CrossRef]
111. Lan, W.R.; Pan, S.; Li, H.Y.; Sun, C.; Chang, X.; Lu, K.; Jiang, C.Q.; Zuo, R.; Zhou, Y.; Li, C.Q. Inhibition of the Notch1 Pathway Promotes the Effects of Nucleus Pulposus Cell-Derived Exosomes on the Differentiation of Mesenchymal Stem Cells into Nucleus Pulposus-Like Cells in Rats. *Stem Cells Int.* **2019**, *2019*, 8404168. [CrossRef]
112. Lu, K.; Li, H.Y.; Yang, K.; Wu, J.L.; Cai, X.W.; Zhou, Y.; Li, C.Q. Exosomes as potential alternatives to stem cell therapy for intervertebral disc degeneration: In-vitro study on exosomes in interaction of nucleus pulposus cells and bone marrow mesenchymal stem cells. *Stem Cell Res. Ther.* **2017**, *8*, 108. [CrossRef] [PubMed]

113. Xia, C.; Zeng, Z.; Fang, B.; Tao, M.; Gu, C.; Zheng, L.; Wang, Y.; Shi, Y.; Fang, C.; Mei, S.; et al. Mesenchymal stem cell-derived exosomes ameliorate intervertebral disc degeneration via anti-oxidant and anti-inflammatory effects. *Free Radic Biol. Med.* **2019**, *143*, 1–15. [CrossRef]
114. Hingert, D.; Ekstrom, K.; Aldridge, J.; Crescitelli, R.; Brisby, H. Extracellular vesicles from human mesenchymal stem cells expedite chondrogenesis in 3D human degenerative disc cell cultures. *Stem Cell Res. Ther.* **2020**, *11*, 323. [CrossRef] [PubMed]
115. Li, M.; Li, R.; Yang, S.; Yang, D.; Gao, X.; Sun, J.; Ding, W.; Ma, L. Exosomes Derived from Bone Marrow Mesenchymal Stem Cells Prevent Acidic pH-Induced Damage in Human Nucleus Pulposus Cells. *Med. Sci. Monit.* **2020**, *26*, e922928. [CrossRef]
116. Li, Z.Q.; Kong, L.; Liu, C.; Xu, H.G. Human Bone Marrow Mesenchymal Stem Cell-derived Exosomes Attenuate IL-1beta-induced Annulus Fibrosus Cell Damage. *Am. J. Med. Sci.* **2020**, *360*, 693–700. [CrossRef]
117. Luo, L.; Jian, X.; Sun, H.; Qin, J.; Wang, Y.; Zhang, J.; Shen, Z.; Yang, D.; Li, C.; Zhao, P.; et al. Cartilage endplate stem cells inhibit intervertebral disc degeneration by releasing exosomes to nucleus pulposus cells to activate Akt/autophagy. *Stem Cells* **2021**, *39*, 467–481. [CrossRef] [PubMed]
118. Luo, L.; Gong, J.; Zhang, H.; Qin, J.; Li, C.; Zhang, J.; Tang, Y.; Zhang, Y.; Chen, J.; Zhou, Y.; et al. Cartilage Endplate Stem Cells Transdifferentiate Into Nucleus Pulposus Cells via Autocrine Exosomes. *Front. Cell Dev. Biol.* **2021**, *9*, 648201. [CrossRef]
119. Tang, S.; Salazar-Puerta, A.; Richards, J.; Khan, S.; Hoyland, J.A.; Gallego-Perez, D.; Walter, B.; Higuita-Castro, N.; Purmessur, D. Non-viral reprogramming of human nucleus pulposus cells with FOXF1 via extracellular vesicle delivery: An in vitro and in vivo study. *Eur. Cell Mater.* **2021**, *41*, 90–107. [CrossRef]
120. Sun, Z.; Zhao, H.; Liu, B.; Gao, Y.; Tang, W.H.; Liu, Z.H.; Luo, Z.J. AF cell derived exosomes regulate endothelial cell migration and inflammation: Implications for vascularization in intervertebral disc degeneration. *Life Sci.* **2021**, *265*, 118778. [CrossRef]
121. Liu, L.T.; Huang, B.; Li, C.Q.; Zhuang, Y.; Wang, J.; Zhou, Y. Characteristics of stem cells derived from the degenerated human intervertebral disc cartilage endplate. *PLoS ONE* **2011**, *6*, e26285. [CrossRef] [PubMed]
122. Chiang, C.K.; Wang, C.C.; Lu, T.F.; Huang, K.H.; Sheu, M.L.; Liu, S.H.; Hung, K.Y. Involvement of Endoplasmic Reticulum Stress, Autophagy, and Apoptosis in Advanced Glycation End Products-Induced Glomerular Mesangial Cell Injury. *Sci. Rep.* **2016**, *6*, 34167. [CrossRef] [PubMed]
123. Song, Y.; Wang, Y.; Zhang, Y.; Geng, W.; Liu, W.; Gao, Y.; Li, S.; Wang, K.; Wu, X.; Kang, L.; et al. Advanced glycation end products regulate anabolic and catabolic activities via NLRP3-inflammasome activation in human nucleus pulposus cells. *J. Cell Mol. Med.* **2017**, *21*, 1373–1387. [CrossRef]
124. Hu, Z.-L.; Li, H.-Y.; Chang, X.; Li, Y.-Y.; Liu, C.-H.; Gao, X.-X.; Zhai, Y.; Chen, Y.-X.; Li, C.-Q. Exosomes derived from stem cells as an emerging therapeutic strategy for intervertebral disc degeneration. *World J. Stem Cells* **2020**, *12*, 803–813. [CrossRef]
125. Cazzanelli, P.; Wuertz-Kozak, K. MicroRNAs in Intervertebral Disc Degeneration, Apoptosis, Inflammation, and Mechanobiology. *Int. J. Mol. Sci.* **2020**, *21*, 3601. [CrossRef] [PubMed]
126. Wang, X.Q.; Tu, W.Z.; Guo, J.B.; Song, G.; Zhang, J.; Chen, C.C.; Chen, P.J. A Bioinformatic Analysis of MicroRNAs' Role in Human Intervertebral Disc Degeneration. *Pain Med.* **2019**, *20*, 2459–2471. [CrossRef]
127. Cheng, X.; Zhang, G.; Zhang, L.; Hu, Y.; Zhang, K.; Sun, X.; Zhao, C.; Li, H.; Li, Y.M.; Zhao, J. Mesenchymal stem cells deliver exogenous miR-21 via exosomes to inhibit nucleus pulposus cell apoptosis and reduce intervertebral disc degeneration. *J. Cell Mol. Med.* **2018**, *22*, 261–276. [CrossRef] [PubMed]
128. Zhang, D.Y.; Wang, Z.J.; Yu, Y.B.; Zhang, Y.; Zhang, X.X. Role of microRNA-210 in human intervertebral disc degeneration. *Exp. Ther. Med.* **2016**, *11*, 2349–2354. [CrossRef]
129. Moen, A.; Jacobsen, D.; Phuyal, S.; Legfeldt, A.; Haugen, F.; Roe, C.; Gjerstad, J. MicroRNA-223 demonstrated experimentally in exosome-like vesicles is associated with decreased risk of persistent pain after lumbar disc herniation. *J. Transl. Med.* **2017**, *15*, 89. [CrossRef]
130. Gu, S.X.; Li, X.; Hamilton, J.L.; Chee, A.; Kc, R.; Chen, D.; An, H.S.; Kim, J.S.; Oh, C.D.; Ma, Y.Z.; et al. MicroRNA-146a reduces IL-1 dependent inflammatory responses in the intervertebral disc. *Gene* **2015**, *555*, 80–87. [CrossRef]
131. Chen, G.; Zhou, X.; Li, H.; Xu, Z. Inhibited microRNA-494-5p promotes proliferation and suppresses senescence of nucleus pulposus cells in mice with intervertebral disc degeneration by elevating TIMP3. *Cell Cycle* **2021**, *20*, 11–22. [CrossRef]
132. Zhu, L.; Shi, Y.; Liu, L.; Wang, H.; Shen, P.; Yang, H. Mesenchymal stem cells-derived exosomes ameliorate nucleus pulposus cells apoptosis via delivering miR-142-3p: Therapeutic potential for intervertebral disc degenerative diseases. *Cell Cycle* **2020**, *19*, 1727–1739. [CrossRef]
133. Wang, W.; Guo, Z.; Yang, S.; Wang, H.; Ding, W. Upregulation of miR-199 attenuates TNF-alpha-induced Human nucleus pulposus cell apoptosis by downregulating MAP3K5. *Biochem. Biophys. Res. Commun.* **2018**, *505*, 917–924. [CrossRef]
134. Wen, T.; Wang, H.; Li, Y.; Lin, Y.; Zhao, S.; Liu, J.; Chen, B. Bone mesenchymal stem cell-derived extracellular vesicles promote the repair of intervertebral disc degeneration by transferring microRNA-199a. *Cell Cycle* **2021**, *20*, 256–270. [CrossRef] [PubMed]
135. Liu, G.; Cao, P.; Chen, H.; Yuan, W.; Wang, J.; Tang, X. MiR-27a regulates apoptosis in nucleus pulposus cells by targeting PI3K. *PLoS ONE* **2013**, *8*, e75251. [CrossRef] [PubMed]
136. Zhang, Q.C.; Hu, S.Q.; Hu, A.N.; Zhang, T.W.; Jiang, L.B.; Li, X.L. Autophagy-activated nucleus pulposus cells deliver exosomal miR-27a to prevent extracellular matrix degradation by targeting MMP-13. *J. Orthop. Res.* **2021**, *39*, 1921–1932. [CrossRef]
137. Zhu, G.; Yang, X.; Peng, C.; Yu, L.; Hao, Y. Exosomal miR-532-5p from bone marrow mesenchymal stem cells reduce intervertebral disc degeneration by targeting RASSF5. *Exp. Cell Res.* **2020**, *393*, 112109. [CrossRef] [PubMed]

138. Zhang, J.; Zhang, J.; Zhang, Y.; Liu, W.; Ni, W.; Huang, X.; Yuan, J.; Zhao, B.; Xiao, H.; Xue, F. Mesenchymal stem cells-derived exosomes ameliorate intervertebral disc degeneration through inhibiting pyroptosis. *J. Cell Mol. Med.* **2020**, *24*, 11742–11754. [CrossRef]
139. Xie, L.; Chen, Z.; Liu, M.; Huang, W.; Zou, F.; Ma, X.; Tao, J.; Guo, J.; Xia, X.; Lyu, F.; et al. MSC-Derived Exosomes Protect Vertebral Endplate Chondrocytes against Apoptosis and Calcification via the miR-31-5p/ATF6 Axis. *Mol. Ther. Nucleic Acids* **2020**, *22*, 601–614. [CrossRef]
140. Song, J.; Chen, Z.H.; Zheng, C.J.; Song, K.H.; Xu, G.Y.; Xu, S.; Zou, F.; Ma, X.S.; Wang, H.L.; Jiang, J.Y. Exosome-Transported circRNA_0000253 Competitively Adsorbs MicroRNA-141-5p and Increases IDD. *Mol. Ther. Nucleic Acids* **2020**, *21*, 1087–1099. [CrossRef]

Article

Effects of Growth Factor Combinations TGFβ3, GDF5 and GDF6 on the Matrix Synthesis of Nucleus Pulposus and Nasoseptal Chondrocyte Self-Assembled Microtissues

Shani Samuel [1,2], Emily E. McDonnell [1,2] and Conor T. Buckley [1,2,3,4,*]

1 Trinity Centre for Biomedical Engineering, Trinity Biomedical Sciences Institute, Trinity College Dublin, The University of Dublin, D02 R590 Dublin, Ireland; samuelsh@tcd.ie (S.S.); mcdonne5@tcd.ie (E.E.M.)
2 Discipline of Mechanical, Manufacturing and Biomedical Engineering, School of Engineering, Trinity College Dublin, The University of Dublin, D02 PN40 Dublin, Ireland
3 Advanced Materials and Bioengineering Research (AMBER) Centre, Royal College of Surgeons in Ireland & Trinity College Dublin, The University of Dublin, D02 PN40 Dublin, Ireland
4 Tissue Engineering Research Group, Department of Anatomy and Regenerative Medicine, Royal College of Surgeons in Ireland, 121/122 St. Stephen's Green, D02 H903 Dublin, Ireland
* Correspondence: conor.buckley@tcd.ie

Abstract: There has been significant interest in identifying alternative cell sources and growth factor stimulation to improve matrix synthesis for disc repair. Recent work has identified nasoseptal chondrocytes (NC) as a possible alternative cell source with significant matrix-forming abilities. While various growth factors such as members of the TGFβ superfamily have been explored to enhance matrix formation, no consensus exists as to the optimum growth factor needed to induce cells towards a discogenic phenotype. This study assessed both nucleus pulposus (NP) and NC microtissues of different densities (1000, 2500 or 5000 cells/microtissue) stimulated by individual or combinations of the growth factors TGFβ3, GDF5, and GDF6. Lower cell densities result in increased sGAG/DNA and collagen/DNA levels due to higher nutrient availability levels. Our findings suggest that growth factors exert differential effects on matrix synthesis depending on the cell type. NP cells were found to be relatively insensitive to the different growth factor types examined in isolation or in combination. Overall, NCs exhibited a higher propensity to form extracellular matrix compared to NP cells. In addition, stimulating NC-microtissues with GDF5 or TGFβ3 alone induced enhanced matrix formation and may be an appropriate growth factor to stimulate this cell type for disc regeneration.

Keywords: microwell; GFD5; GDF6; TGFβ3; microtissues; intervertebral disc

Citation: Samuel, S.; McDonnell, E.E.; Buckley, C.T. Effects of Growth Factor Combinations TGFβ3, GDF5 and GDF6 on the Matrix Synthesis of Nucleus Pulposus and Nasoseptal Chondrocyte Self-Assembled Microtissues. *Appl. Sci.* **2022**, *12*, 1453. https://doi.org/10.3390/app12031453

Academic Editor: Benjamin Gantenbein

Received: 13 December 2021
Accepted: 27 January 2022
Published: 29 January 2022

1. Introduction

Cell-based therapies may hold significant potential as a treatment strategy for the repair and regeneration of the intervertebral disc (IVD). However, it remains challenging to identify an appropriate cell source that can be obtained with minimal donor site morbidity and produce a matrix with high levels of proteoglycan containing predominately collagen type II and low levels of collagen type I. The most commonly explored cell types for IVD regeneration include mesenchymal stem cells (MSCs) [1,2], articular chondrocytes [3] and disc-derived cells [4]. Studies have demonstrated that reimplantation of extracted NP cells can retard degenerative changes, and the injection of autologous nucleus pulposus (NP) cells has been clinically tested in humans with positive outcomes [5]. DiscGenics Inc. (Salt Lake City, Utah) is actively investigating their propriety technology IDCT (injectable discogenic cell therapy) which contains a mixture of progenitor cells derived from allogeneic "discogenic" cells and a viscous biomaterial [6]. However, the use of NP cells is limited due to the matrix-forming capacity of expanded NP cells derived from degenerated tissue [7] and the number of nucleus pulposus (NP) cells that can be isolated from a degenerated disc is insufficient [8] to meet the requirements for successful treatment without significant culture expansion.

In contrast, stem cells can be obtained from various sources (bone marrow or adipose tissue) and have been shown to survive and proliferate after implantation into the disc [9]. However, cell leakage at the injection site has been shown to result in osteophyte formation [10]. Mature chondrocytes produce a matrix similar to NP cells with nasoseptal chondrocytes (NCs) being recently explored as an alternative non-disc cell source for IVD regeneration. NCs have a higher cell yield and sulphated glycosaminoglycan (sGAG) synthesis [11] when compared to articular chondrocytes (ACs) and MSCs [12]. Recently, Borrelli et al. also demonstrated that NCs resulted in higher matrix synthesis levels in comparison to NP cells [13]. The use of NCs for IVD regeneration may present many advantages compared to NP cells, including a high cell yield [11], the ability to produce tissue in an age-independent manner [14,15], enhanced matrix synthesis [13] and can be obtained with minimal donor site morbidity facilitating their autologous use. These intrinsic properties make NCs an attractive non-disc cell source for IVD regeneration.

Growth factors regulate IVD homeostasis, including extracellular matrix (ECM) synthesis and degradation, cell differentiation, or apoptosis [16]. Researchers have widely investigated the potential of various growth factor-mediated induction of cells to mimic the expression profile resembling native disc tissue. Studies have shown that the activation of transforming growth factor beta (TGFβ) signalling pathways delays IVD degeneration by increasing ECM synthesis [17,18] and hence members of the TGFβ superfamily may be potential candidates for driving discogenic differentiation and phenotype. TGFβ3 has been shown to maintain the phenotype of disc cells in organ culture [19] and when encapsulated with MSCs, they induced IVD regeneration in vivo [20]. Growth differentiation factor 5 (GDF5) and GDF6 are members of the bone morphogenetic protein (BMP) family. GDF5 and GDF6 play important roles in the development of bones, limb joints, skull and axial skeleton [21] and are also expressed in developing cartilage, tendons and ligaments [22,23]. In GDF5/6-knockout mouse models, the vertebral column showed severe lateral curvature and reduced staining of the IVDs indicating lower proteoglycan content [22]. These results suggest that GDF5 and GDF6 are required for normal development and maintenance of the IVD. In pellet culture, GDF5 results in reduced expression of the catabolic enzyme MMP13 in human chondrocytes [24]. There is also evidence to suggest that GDF-5 confers protection against NP cell apoptosis, promotes the synthesis of the main components of the extracellular matrix and can inhibit the activation of the NF-κB signalling pathway, thereby down-regulating the expression of inflammatory cytokines [25]. Inflammatory cytokines play an important role in the pathogenesis of disc degeneration by promoting matrix breakdown and recruitment of immune cells [26]. GDF6 supplementation has been shown to increase both proteoglycan and collagen production in 3D alginate bead cultures of human NP and AF cells [27]. Clarke et al. previously tested the individual effects of TGFβ3, GDF5, and GDF6 on discogenesis of bone marrow-derived MSCs and adipose-derived stem cells (ADSCs) reporting that GDF6 induced differentiation of these cell types towards an NP-like phenotype to a greater extent compared to other growth factors [28]. Furthermore, the authors found optimal expression of genes COL2 and ACAN when stimulated with 10 ng/mL TGFβ3 and 100 ng/mL GDF5 and GDF6. However, they did not investigate the effect of combining these growth factors. The overall objective of this study was to assess and compare the extent of matrix deposited by NP and NC microtissues stimulated by individual or combinations of the growth factors TGFβ3, GDF5, and GDF6.

2. Materials and Methods

2.1. Nucleus Pulposus and Nasoseptal Chondrocyte Isolation and Expansion

All tissue was sourced from a local abattoir and dissected within 24 h. NP cells were isolated from the IVDs of porcine spines (3 female donors; 4 months old). Briefly, NP tissues were harvested aseptically from the central nucleus pulposus of the IVD avoiding the annulus fibrosus region, washed with phosphate-buffered saline (PBS) and minced. Tissue fragments were placed in T-25 flasks containing Low Glucose-Dulbecco's Modified Eagle Medium (LG-DMEM) with 10% foetal bovine serum (FBS) and penicillin (100 U/mL)-streptomycin (100 μg/mL) (PenStrep) and cultured in a humidified atmosphere at 37 °C and 5 %CO_2. After ~7 days, cells had migrated from the tissue fragments, the flasks were washed with PBS and the cells were expanded to 80% confluence. NCs were isolated from porcine nasal septa, washed with PBS and minced. To isolate NC cells, minced tissue was digested in serum free LG-DMEM containing PenStrep and 3000 U/mL collagenase type II (Gibco, Invitrogen, Ireland) at a ratio of 10 mL/g of minced tissue. Digestion was performed under constant rotation for 3 h at 37 °C and subjected to physical agitation using a tissue dissociator (GentleMACS^{TM}, Miltenyi Biotech, Surrey, UK) as previously described [29]. Digested tissue/cell suspensions were passed through a 40 μm cell strainer to remove tissue debris. Cell yield and viability in both cases were determined with a hemocytometer and trypan blue exclusion. Cells were seeded at an initial density of 5×10^3 cells/cm^2 in T-175 flasks in LG-DMEM supplemented with 10% FBS and PenStrep. Cultures were expanded to passage two in a humidified atmosphere at 37 °C and 5 %CO_2.

2.2. Fabrication of PDMS Microwell Moulds and Microtissue Formation

Concave PDMS (polydimethylsiloxane) microwells were employed to generate self-assembled microtissues using lower numbers of cells than traditional pellet cultures or alginate bead models using a similar approach as described previously [30]. The microwell layout was designed using SolidWorks software (Solid Solutions Management Ltd, Dublin, Ireland). Moulds were designed with 100 microwells (10 × 10 array) containing 500 μm hemispherical concave microwells, with 1390 μm centre–centre distance and a depth of 1750 μm (Figure 1). A Formlabs Form 2 printer (Formlabs GmbH, Berlin, Germany) was used to 3D print the clear resin master stamps. After printing, the master stamps were washed in isopropyl alcohol for 20 min to remove excess residue and subsequently cured under an ultraviolet lamp (4 W, Uvitec, Cambridge, UK) for 1 h. 3D printed stamps were inserted into 12-well plates containing approximately 1 mL of PDMS (Sylgard® 184, Sigma-Aldrich, Arklow, Ireland), degassed in a vacuum oven and cured at 80 °C for 4 h. After cooling, the master stamps were removed carefully with the use of a tweezers. The PDMS microwell moulds formed were washed with ethanol and phosphate-buffered saline (PBS, pH 7.4), sterilised by UV-irradiation for 3 h and allowed to dry overnight in a laminar hood. The morphology and the diameter of the PDMS microwells formed were characterised by scanning electron microscopy (SEM) (SEM, SUPRA 35 V P, Carl Zeiss, Oberkochen, Germany). For microtissue formation, the moulds were placed in 12-well plates and seeded with different densities of NP and NC cells (1000, 2500, and 5000 cells/microwell or 100,000, 250,000, 500,000 cells/mould) and centrifuged at 300× g for 3 min to initiate microtissue formation. The plates were then cultured in a humified atmosphere at 37 °C and 5 %O_2 in 1 mL of media supplemented with or without growth factors as described below. Preliminary experiments investigated the formation of microtissues with less than 1000 cells/microtissue. However, these were found to be inconsistent and lacked cohesion and integrity, making them unsuitable for further experiments.

2.3. Growth Factor Stimulation

Microtissues were cultured in 1 mL of LG-DMEM with PenStrep, 1.5 mg/mL bovine serum albumin (BSA; Sigma-Aldrich), 1× insulin-transferrin-sodium selenite (ITS; Sigma-Aldrich), 40 μg/mL L-proline, 100 nM Dexamethasone, 50 μg/mL L-ascorbic acid 2-phosphate, 4.7 μg/mL linoleic acid and were either supplemented with no growth factor (control) or with 10 ng/mL TGFβ3 [31]. Microtissues were maintained at 37 °C for 7 days under low oxygen (5 %O_2) conditions. After 7 days of culture, microtissue size was evaluated through image analysis using Image J software (NIH, Bethesda, MA, USA). Initial results showed that lower cell density microtissues (1000 cells/microtissue) resulted in higher matrix production. Hence, this cell density was chosen for further experiments and they were individually supplemented with either 100 ng/mL of GDF5, 100 ng/mL of GDF6 [28], 10 ng/mL TGFβ3 + 100 ng/mL GDF5, 10 ng/mL TGFβ3 + 100 ng/mL GDF6 or 10 ng/mL TGFβ3 + 100 ng/mL GDF5 + 100 ng/mL GDF6.

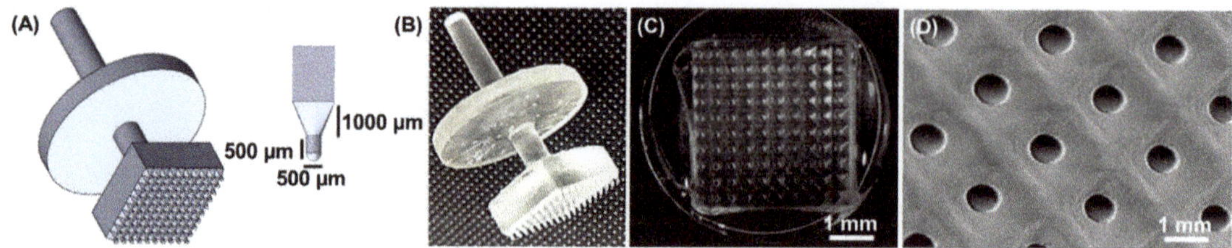

Figure 1. Fabrication of PDMS microwells for microtissue formation. (**A**) Design of the master stamp with negative pattern containing 10 × 10 microwells and the dimensions of the individual microwell. (**B**) 3D printed master stamp. (**C**) PDMS moulds with a 10 × 10 array for microtissue culture. (**D**) Microwells exhibited uniformity in size and shape as confirmed by SEM imaging.

2.4. In-Silico Modelling of Microtissue Nutrient Microenvironment

The in silico nutrient model of microtissues was created using COMSOL Multiphysics 5.6 (COMSOL Ltd., Cambridge, UK). Oxygen concentration at the microtissue boundary was dependent on multiple local environmental factors such as external level at the air/media interface, diffusion rate within the media, and the volume of media used. Glucose concentration and pH level at the microtissue boundary was dependent on the initial concentration of the media and the volume of media. Therefore, the in silico model was based on the in vitro geometry of the media-filled microwell containing an idealised radially symmetric microtissue with an initial cell seeding density of 1000, 2500 or 5000 cells. Since oxygen kinetics occur on a faster timescale than cell cycle kinetics the radius of the microtissue was assumed to be constant. The steady-state nutrient microenvironment was governed by coupled reaction-diffusion equations as described previously for disc cells [32]. Briefly, the metabolic rates were modelled as being dependent on local oxygen and pH levels by employing Michaelis–Menten equations derived and published previously [32–35]. Results for oxygen, glucose and pH levels were predicted and displayed as concentration contour plots through the midplane of the microtissues and graphically as a function of normalised radial distance through the microtissues.

2.5. Live/Dead Analysis

Cell viability was assessed using the LIVE/DEAD® viability/cytotoxicity assay kit (Invitrogen, Biosciences, Dublin, Ireland). Briefly, microtissues were washed with PBS and incubated in live/dead solution containing 2 µM calcein AM (live cell membrane, abs/em = 494/517 nm) and 4 µM ethidium homodimer-1 (dead cell DNA, ex/em = 528/617 nm; both from Cambridge Bioscience, Cambridge, UK) in PBS for 1 h. Samples were then washed in PBS, imaged with a Leica SP8 scanning confocal microscope at 515 and 615 nm channels and assessed using Leica Application Suite X (LAS X) Software (Version 3.5.5.19976).

2.6. Quantitative Biochemical Analysis

After 7 days in culture, microtissues were flushed from the moulds with PBS and frozen at −80 °C for further analysis. Microtissues were digested with 3.88 U/mL papain in 0.1 M sodium acetate, 5 mM L-cysteine-HCl and 0.05 M ethylenediaminetetraacetic acid (EDTA) (pH 6.0) (all from Sigma-Aldrich) at 60 °C under constant rotation for 18 h. DNA content was quantified using the Quant-iT PicoGreen® dsDNA (Invitrogen) assay. sGAG content was quantified using the dimethylmethylene blue dye-binding assay at pH 1.35 (Blyscan, Biocolor Ltd., Carrickfergus, Northern Ireland), with a chondroitin sulphate standard. Total collagen content was determined by measuring the hydroxyproline content. Samples were hydrolysed at 110 °C for 18 h in 12 M HCl, assayed using a chloramine-T assay [36] and the collagen content determined using a hydroxyproline:collagen ratio of 1:7.69. Samples of media supernatants were also analysed for both sGAG and collagen content. DNA and sGAG contents were normalised based on the mass of all the microtissues (~100) from the entire microwell. The sGAG/collagen ratio was determined by dividing sGAG (μg) by collagen (μg).

2.7. Histology and Immunohistochemistry

Microtissues were treated with 4% paraformaldehyde (4 °C, 12 h) and washed in PBS. Microtissues were encapsulated in 2% agarose (Sigma-Aldrich) to facilitate handling, transferred to a cassette, dehydrated in a series of graded alcohols and finally wax embedded. Sections of 5 μm were stained with 1% alcian blue 8GX in 0.1 M HCl to assess sGAG deposition and with picrosirius red to evaluate collagen distribution. Collagen types I and II were assessed using immunohistochemistry techniques. Sections were treated with chondroitinase ABC (37 °C, 1 h) (Sigma-Aldrich), and non-specific sites were blocked using 5% BSA. Sections were incubated at 4 °C overnight with collagen type I (Abcam, Cambridge, UK) or collagen type II (Santa Cruz, Heidelberg, Germany) primary antibodies. The secondary antibody (Anti-Mouse igG biotin conjugate, Sigma-Aldrich) was applied for 1 h followed by incubation (45 min) with ABC reagent (Vectastain PK-400, Vector Labs, 2BScientific Ltd., Oxfordshire, UK). DAB peroxidase (Vector Labs, UK) was used as a developer.

2.8. Statistical Analysis

Statistical analysis was performed using GraphPad Prism (version 9.3.1, GraphPad Software, San Diego, CA, USA) software with 3–4 samples analysed for each experimental group. Two-way ANOVA was used for analysis of variance with Tukey's multiple comparisons test to compare between groups. Three technical replicates from three different porcine donors (biological replicates) were analysed. Results are displayed as mean ± standard deviation, with significance accepted at a level of $p < 0.05$.

3. Results

3.1. Design and Fabrication of PDMS Microwells

The master stamp prototype designed using SolidWorks software (Figure 1A) was fabricated using a Form 2 printer (Figure 1B). The printed stamps had a smooth and clear surface and were used to form the negative PDMS moulds. The PDMS moulds formed had 100 microwells (10 × 10 array) with 500 μm hemispherical concave microwells, with 1390 μm centre-centre distance and depth of 1750 μm (Figure 1C). The microwells had a uniform size with an average diameter of 504 ± 24 μm based on SEM images (Figure 1D).

3.2. Morphology and Viability of Microtissues

In terms of cell viability, results revealed high viability in low cell density microtissues with more dead cells being detected in the microtissues formed using higher cell numbers (Figure 2A). Moreover, the microtissues in the control group were loosely packed, exhibiting an irregular surface with loose arrangement of cells at the peripheral layers while the microtissues formed in the presence of TGFβ3 exhibited a smooth surface with compact organisation of the interior cells. The low cell density microtissues in the control group were also easily dissociated during the flushing process. The size of the microtissues increased with increasing cell numbers in both control and TGFβ3 treated groups for both NP and NC microtissues (Figure 2B). As expected, DNA content increased with higher content observed in higher cell density microtissues. NC microtissues exhibited increased DNA content (which indicates higher cell content) compared to NP microtissues, with a significant difference observed for 2500 and 5000 cells/microtissue cultured in the presence of TGFβ3 (Figure 2C).

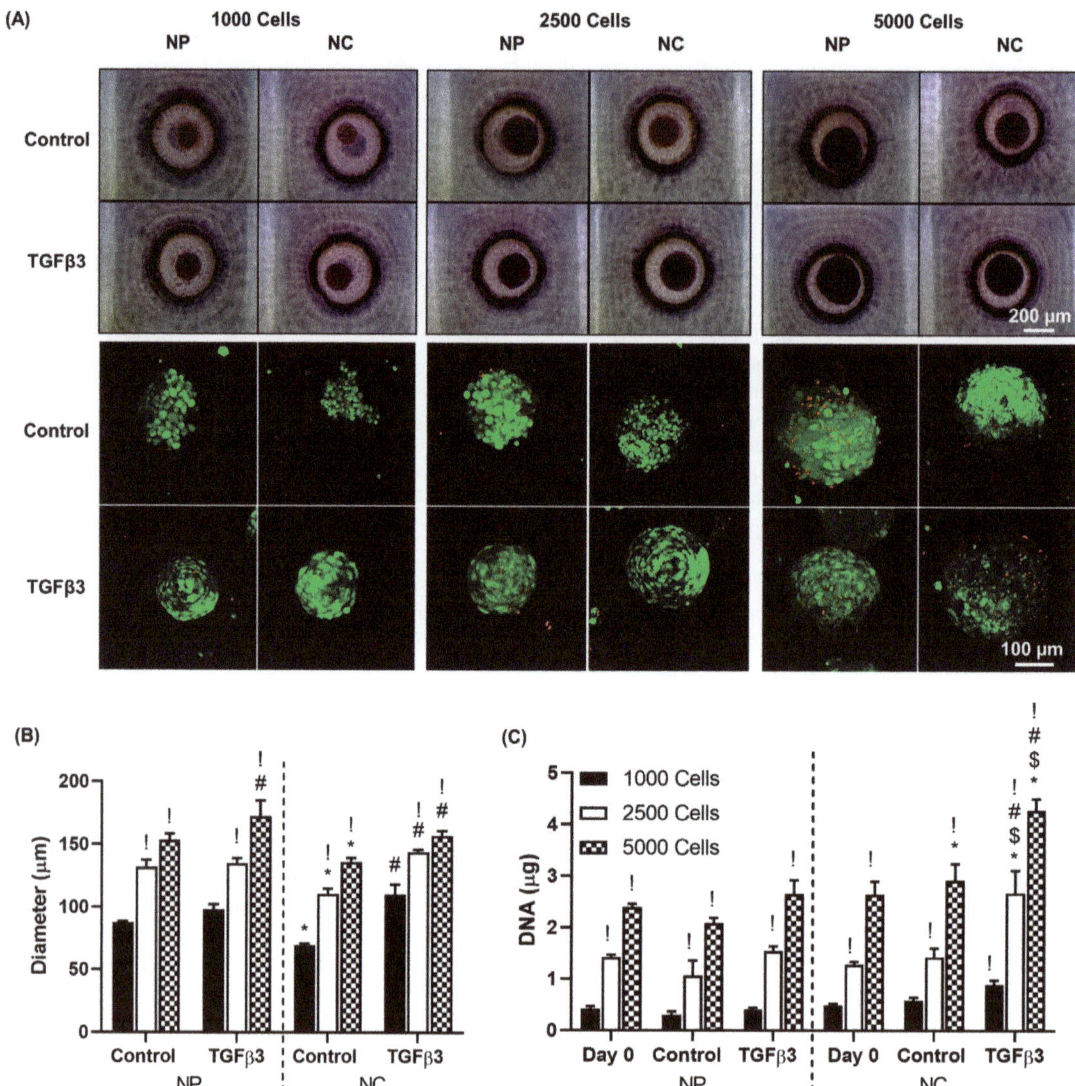

Figure 2. Morphology and viability of microtissues formed using nucleus pulposus (NP) and nasoseptal chondrocyte (NC) cells. (**A**) Representative brightfield and fluorescent Live/Dead® images demonstrating the morphology and viability of microtissues in microwells seeded with 1000, 2500 and 5000 cells/microwell, after 7 days (**B**) Diameter of microtissues were analysed using ImageJ (**C**) DNA content (μg) was determined in digested microtissues from moulds at each timepoint (day 0 or day 7). Data represent mean ± SD of at least three independent experiments performed in triplicate. ! ($p < 0.05$) indicates significance compared to 1000 cells/microtissue. # ($p < 0.05$) indicates significance compared to the control group for the same cell number. $ ($p < 0.05$) indicates significance compared to day 0 for the same cell number. * ($p < 0.05$) indicates significance compared to the NP microtissues for the same experimental group and cell number.

3.3. ECM Synthesis and Optimising Cell Density of TGFβ3 Stimulated Microtissues

Total sGAG content increased with increasing cell densities, with NC microtissues achieving significantly higher sGAG levels for all cell densities compared to NP microtissues cultured in the presence of TGFβ3 (Figure 3A). sGAG normalised to DNA showed increased sGAG/DNA for lower cell density microtissues and the NC 1000 cells/microtissue in TGFβ3 achieved a significantly higher level compared to NP 1000 cells/microtissue (Figure 3B). A similar trend was observed for total collagen content with higher cell density microtissues resulting in higher collagen accumulation (Figure 3C). Interestingly, when normalised to DNA content, NP 1000 cells/microtissue in TGFβ3 was significantly higher compared to NC 1000 cells/microtissue in TGFβ3 (Figure 3D). Strong alcian blue staining in lower cell density NC microtissues in TGFβ3 confirmed the higher synthesis of sGAG (Figure 3E). Picrosirius red staining for collagen also corroborated the collagen biochemical data with dense staining observed in NP 1000 cells/microtissue supplemented with TGFβ3. Comparatively, low cell density microtissues resulted in better ECM synthesis on a per cell basis and in silico modelling was performed to determine the nutrient microenvironment within the microwells to elucidate how this may be influencing matrix synthesis.

3.4. In-Silico Modelling of Microtissue Nutrient Microenvironment

An in silico modelling analysis was performed based on the size of the microtissues for each of the cell densities investigated. As expected, an inverse relationship between metabolite concentration and cell density was observed and correlated well with the biochemical observations, whereby higher ECM on a per cell basis was observed with higher oxygen, glucose and pH levels. In terms of oxygen, minimum levels of 2.7 $\%O_2$, 1.8 $\%O_2$ and 0.9 $\%O_2$ for 1000, 2500 and 5000 cells/microwell were observed, respectively (Figure 4A,B). For glucose, minimum levels of 4.6 mM, 4.1 mM and 3.4 mM (Figure 4C,D) and for pH, 7.3, 7.2 and 7.0 were observed for 1000, 2500 and 5000 cells/microwell respectively (Figure 4E,F). Minimal gradient effects from the centre to the periphery of microtissues were observed.

3.5. Viability and ECM Synthesis of Microtissues in the Presence of Different Growth Factors and Combinations

Based on the observations of lower cell density microtissues resulting in higher cell viability and matrix production, a cell density of 1000 cells/microtissue was chosen for further experiments. Microtissues formed by NCs were more compact compared to NP microtissues. No significant difference in viability was observed among the microtissues cultured in the presence of various growth factors, with high cell viability observed on day 7 (Figure 5A). Weak alcian blue staining for sGAG was observed in NP microtissues supplemented with GDF5 and GDF6 growth factors in isolation. Increased positive staining was observed by co-stimulation with TGFβ3. For NC microtissues, GDF5 had a positive effect with minimal staining observed for GDF6. TGFβ3 co-stimulation resulted in intense staining and was comparatively stronger than NP microtissues (Figure 5B). Both NP and NC microtissues stained positive for collagen in the presence of GDF5, with higher deposition of collagen observed in the presence of GDF6 and when co-stimulated with TGFβ3 (Figure 5C).

3.6. Quantitative Biochemical Analysis of Microtissues in the Presence of Different Growth Factors

Both cell types responded positively to growth factor treatment relative to the no growth factor control (red solid line). For NP microtissues, no significant differences in DNA content were observed after 7 days irrespective of the growth factor type or combination. In contrast, a higher DNA content was observed in NC microtissues when cultured in the presence of GDF5 and all other combinations of growth factors relative to GDF6 (Figure 6A). Total sGAG was highest in NP microtissues stimulated with TGFβ3 + GDF5, although there were no significant differences between the groups investigated. Total sGAG levels in the NC microtissues were similar for all growth factor groups and combinations and were significantly higher compared to GDF6. Overall, sGAG levels for

NC microtissues were significantly higher compared to NP microtissues (Figure 6B). When normalised by DNA content (sGAG/DNA), no differences were found between growth factor groups for a specific cell type, although NC microtissues still exhibited higher content compared to NP cells (Figure 6C).

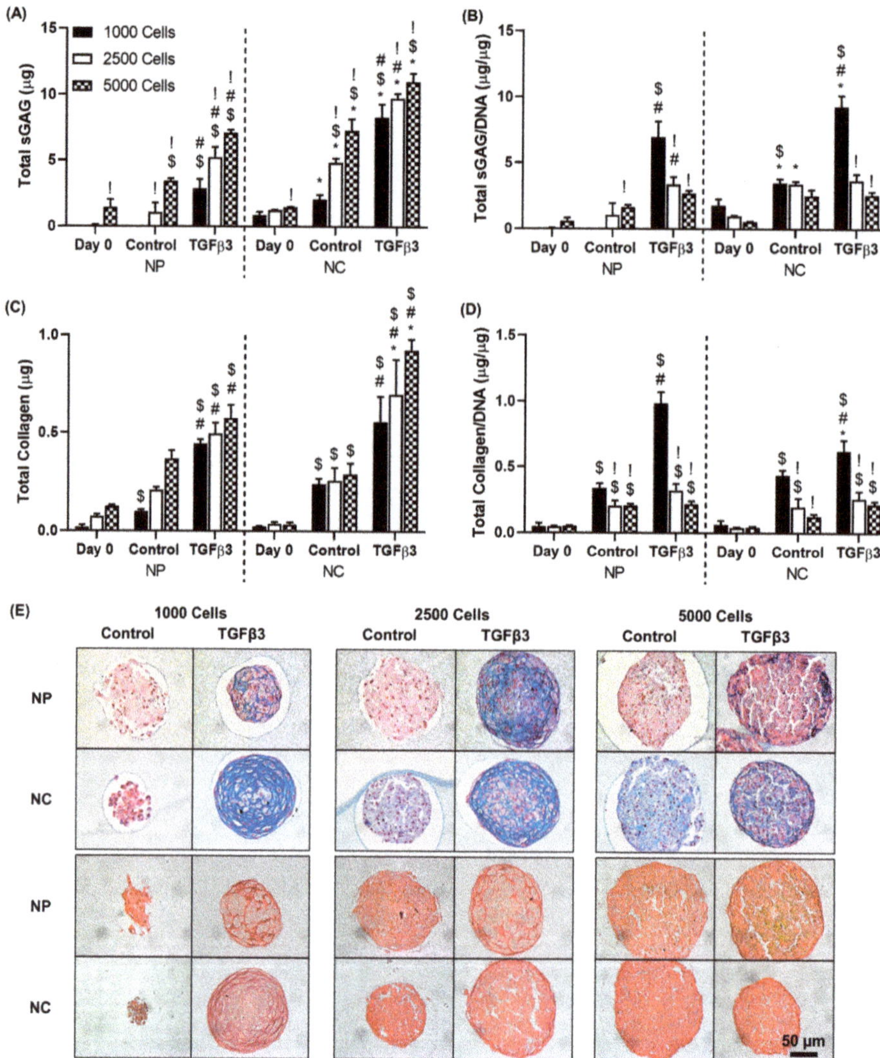

Figure 3. Biochemical and histological staining for sGAG and collagen of nucleus pulposus (NP) and nasoseptal chondrocyte (NC) microtissues stimulated with TGFβ3 on day 7. (**A**) Total sGAG (μg) content (**B**) Total sGAG/DNA (**C**) Total Collagen (μg) (**D**) Total Collagen/DNA. Data represent mean ± SD of at least three independent experiments performed in triplicate. ! ($p < 0.05$) indicates significance compared to 1000 cells/microtissue. # ($p < 0.05$) indicates significance compared to the control for the same cell number on day 7. $ ($p < 0.05$) indicates significance compared to day 0 for the same cell number. * ($p < 0.05$) indicates significance compared to the NP microtissues for the same experimental group and cell number. (**E**) Histological evaluation with alcian blue staining and picrosirius red to identify sGAG collagen content, respectively.

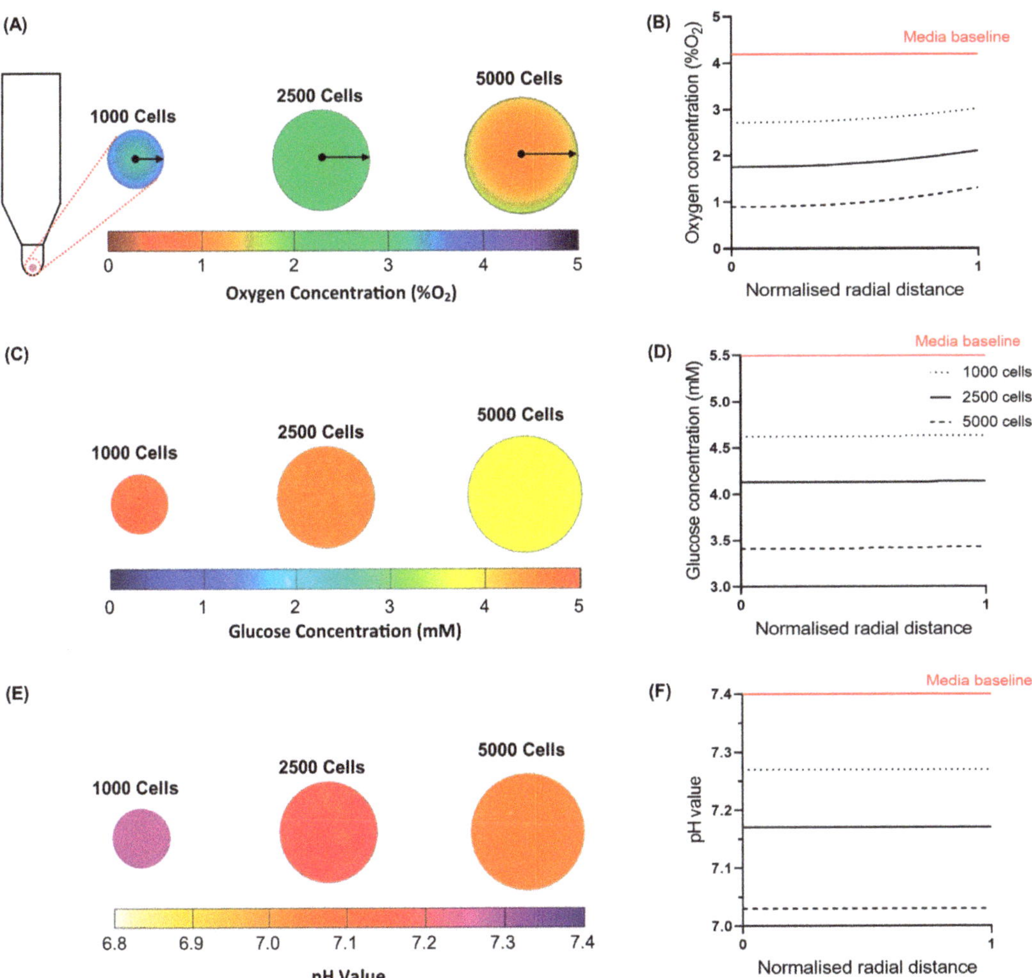

Figure 4. In silico modelling of the nutrient microenvironment within microtissues as a function of cell density. Predicted contour plots and radial profile through the midplane containing 1000, 2500 and 5000 cells for (**A**,**B**) oxygen (%O_2) (**C**,**D**) glucose (mM) and (**E**,**F**) pH. The radial distance was normalised by the radius of the microtissue for each cell density.

In terms of total collagen content, for NP microtissues, a synergistic effect was observed through co-stimulation of growth factors (TGFβ3 + GDF5, TGFβ3 + GDF6 and TGFβ3 + GDF5 + GDF6) when compared to GDF5 and GDF6 alone. A similar synergistic effect was observed for NC microtissues when cultured in the presence of TGFβ3 + GDF5. However, the collagen content in NC microtissues treated with TGFβ3 + GDF6 and TGFβ3 + GDF5 + GDF6 were significantly higher than those treated with GDF5 alone. NC microtissues cultured in the presence of GDF6, TGFβ3 + GDF5 and TGFβ3 + GDF5 + GDF6 exhibited significantly higher collagen content compared to NP microtissues exposed to similar conditions (Figure 6D). Total collagen/DNA was significantly higher in NP microtissues cultured in the presence of TGFβ3 + GDF6, and no significant difference was observed when compared to TGFβ3 (red dashed line). However, for NC microtissues, GDF6-treated groups exhibited significantly higher concentrations compared to TGFβ3 (red dashed line), GDF5 and TGFβ3 + GDF5 + GDF6 treated groups (Figure 6E). In terms of sGAG/collagen level, which is a surrogate measure of NP-like matrix, with a higher ratio being desirable, no significant difference—irrespective of the growth factor treatment—was observed for NP microtissues. NC microtissues treated with GDF5 showed significantly higher concentrations compared to GDF6, TGFβ3 + GDF5 and TGFβ3 + GDF6. They also exhibited higher ratios compared to NP microtissues when cultured in the presence of TGFβ3 (red dashed line), GDF5, TGFβ3 + GDF5, TGFβ3 + GDF6 and TGFβ3 + GDF5 + GDF6 growth factor combinations (Figure 6F).

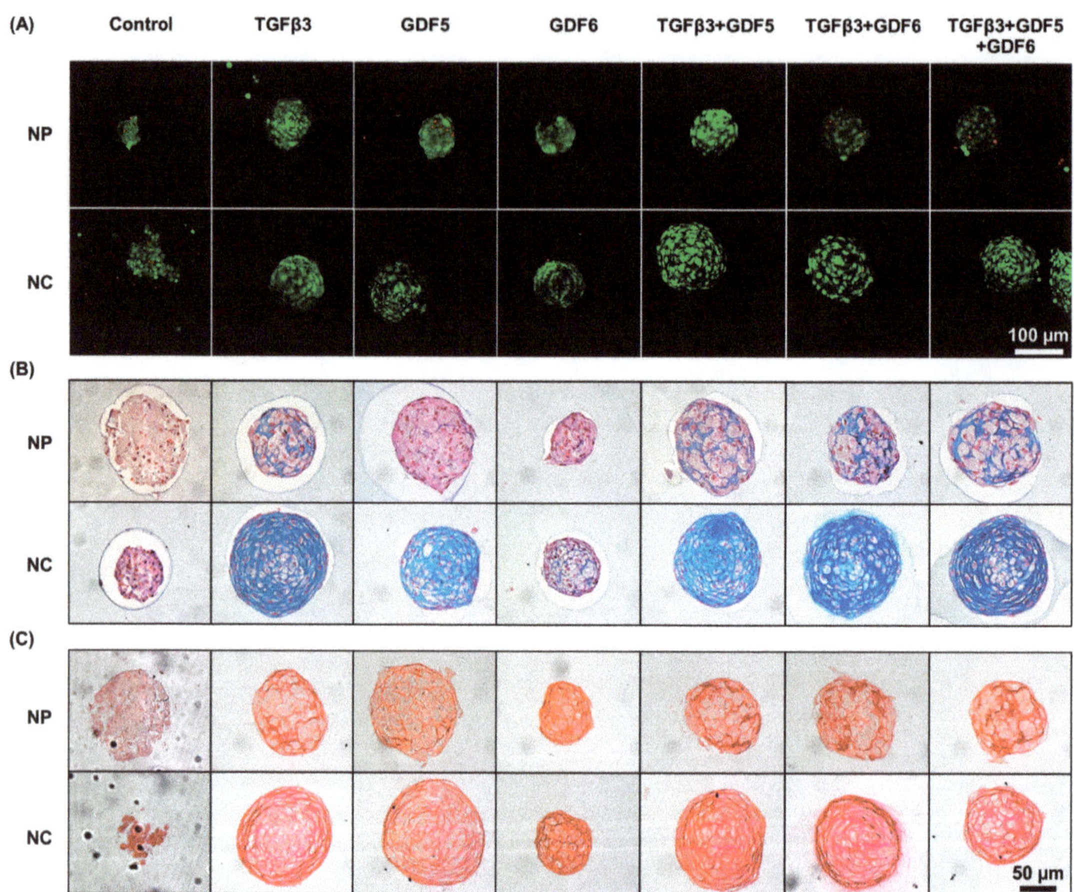

Figure 5. Live/Dead® and histological staining of nucleus pulposus (NP) and nasoseptal chondrocyte (NC) microtissues (1000 cells/microtissue) treated with no growth factor (control) or stimulated with TGFβ3, GDF5, GDF6, TGFβ3 + GDF5, TGFβ3 + GDF6, TGFβ3 + GDF5 + GDF6 on day 7. (**A**) Live/Dead® fluorescent images demonstrating cell viability. Histological evaluation with (**B**) alcian blue staining to identify sGAG and (**C**) picrosirius red to detect collagen.

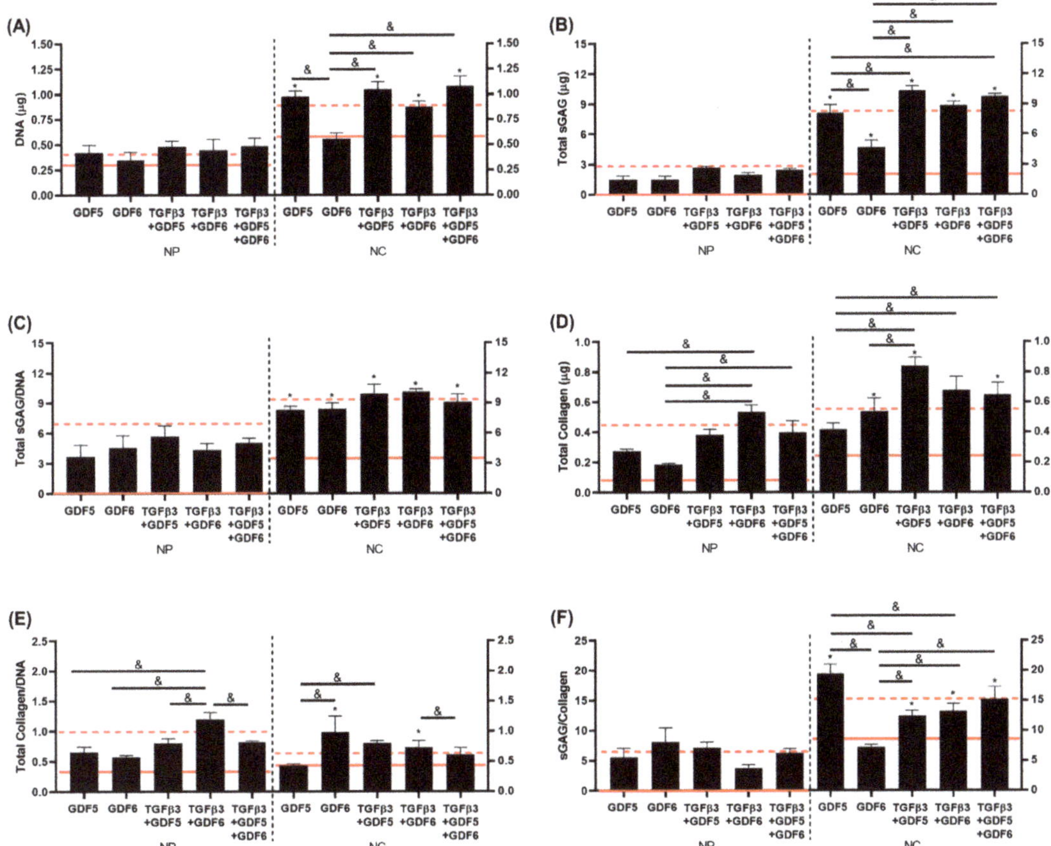

Figure 6. Biochemical analysis of nucleus pulposus (NP) and nasoseptal chondrocyte (NC) microtissues stimulated with GDF5, GDF6, TGFβ3 + GDF5, TGFβ3 + GDF6, TGFβ3 + GDF5 + GDF6 on day 7. (**A**) DNA (µg), (**B**) Total sGAG (µg), (**C**) Total sGAG/DNA, (**D**) Total Collagen (µg), (**E**) Total Collagen/DNA and (**F**) sGAG/Collagen ratio. & ($p < 0.05$) indicates significance compared to other growth factors in the same experimental group, * ($p < 0.05$) indicates significance compared to the NP microtissues for the same growth factor group. Solid red line indicates no growth factor treatment group (control); red dashed line indicates TGFβ3 stimulation group.

3.7. Immunostaining for Collagen Type of Microtissues in the Presence of Different Growth Factors

Irrespective of the growth factor used, immunostaining showed limited presence of collagen type I in both NP and NC microtissues (Figure 7A). NP microtissues also showed weak positive staining for collagen type II while NC microtissues stimulated with TGFβ3, GDF5 and TGFβ3 + GDF5 growth factors had higher and more dispersed collagen type II content (Figure 7B).

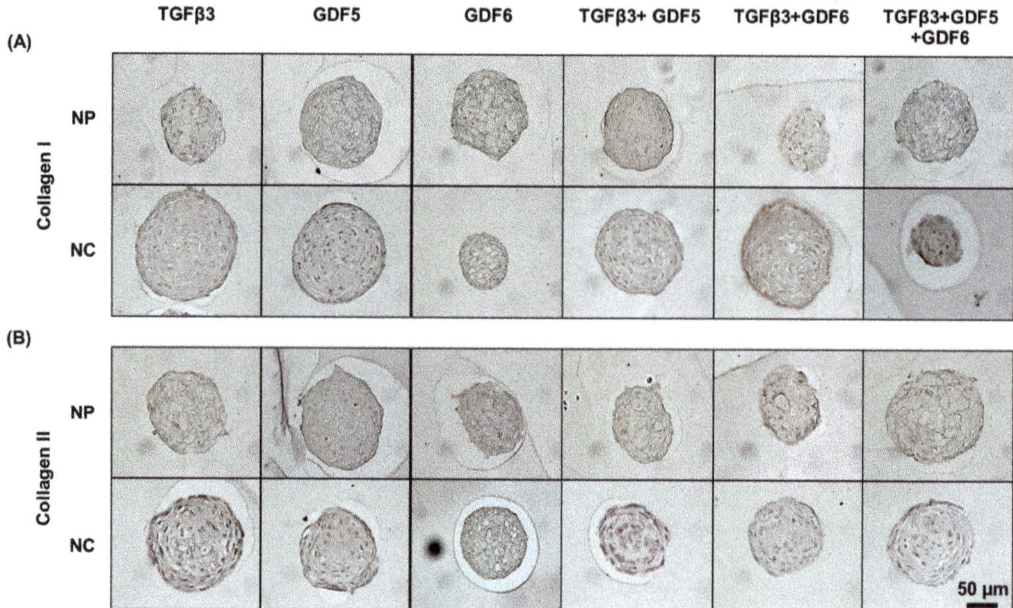

Figure 7. Immunostaining for collagen types I and II of nucleus pulposus (NP) and nasoseptal chondrocyte (NC) microtissues stimulated with TGFβ3, GDF5, GDF6, TGFβ3 + GDF5, TGFβ3 + GDF6, TGFβ3 + GDF5 + GDF6 on day 7. (**A**) Collagen type I and (**B**) Collagen type II.

4. Discussion

In this study, a microwell array system was used to promote the formation of self-assembled homogenous microtissues using two different cell types, nucleus pulposus cells and nasoseptal chondrocytes, and were subsequently assessed in response to growth factor stimulation. Growth factors TGFβ3, GDF5, GDF6 and a combination of these growth factors were assessed for their ability to induce matrix synthesis. The choice of growth factors selected was based on previous research in the field. For example, TGFβ3 is known to support the maintenance of the phenotype of disc cells in a rat lumbar organ culture [19] and MSCs supplemented with TGFβ3 support their differentiation towards an NP-like phenotype [37]. In addition, murine IVD explants treated exogenously with TGFβ3 have been shown to result in an up-regulation of aggrecan [38]. Studies have also shown that cells primed with TGFβ3 promoted higher sGAG and collagen synthesis [31,39] and also mitigated the detrimental effect of the harsh acidic microenvironment [40]. GDF5 (also known as cartilage-derived morphogenetic protein, CDMP1) and GDF6 (CDMP2) are members of the bone morphogenetic protein (BMP) family and are key regulators of cellular condensation and chondrogenesis [41]. GDF5 binds with higher affinity to BMP receptor type-1B (BMPR-IB), which controls the primary stages of cellular condensation and promotes cartilaginous tissue formation [42,43]. GDF6 has been shown to promote chondrogenesis and it positively regulates growth and maintenance of articular cartilage [44]. Both GDF5 and GDF6 have been shown to induce the expression of NP associated genes in MSCs and adipose-derived stem cells (ASCs) [45] with a healthy NP phenotype exhibiting stabilised expression of hypoxia inducible factor HIF-1α, the glucose transporter glut-1, the PG aggrecan (ACAN), type II collagen (COL2A), the signalling factor sonic hedgehog (SHH), the transcription factor Brachyury [T], keratins KRT18, KRT19, carbonic anhydrase CA12, and CD24 [46].

We selected the growth factor concentrations of TGFβ3, GDF5, and GDF6, based on previous work stimulating bone marrow-derived MSCs and adipose-derived stem cells

(ADSCs) which reported optimal expression of genes COL2 and ACAN when stimulated with 10 ng/mL TGFβ3 and 100 ng/mL GDF5 and GDF6 [28].

This study confirms that the initial cell seeding density of microtissues plays an important role in regulating ECM deposition. Although the amount of sGAG and collagen produced were highest in high cell density microtissues, sGAG and collagen synthesised on a per cell basis (per DNA) were highest in low density microtissues (1000 cells/microtissues). Based on the in silico models, higher cell density microtissues experienced lower nutrient levels and this in turn could alter the metabolism of the cells [32,47], and inhibit or impede matrix synthesis [39,48]. Overall, higher levels of matrix synthesis were observed for NC microtissues compared to NP microtissues, and it was significantly pronounced for NC microtissues stimulated with GDF5 or in combination with TGFβ3. In agreement with previous studies, microtissues stimulated with TGFβ3 resulted in increased proteoglycan synthesis compared to the controls [49,50]. TGFβ3 is widely known to modulate cell proliferation, differentiation and ECM synthesis and in particular supports chondrogenic differentiation of various cell types including mesenchymal stem cells [37,51–54], through the activation of the Wnt signalling pathway [55,56]. It has also been reported that Smad2/3 cooperatively is one of the important signalling pathways stimulated by TGFβ3 that helps in developing and maintaining a chondrocytic phenotype [57].

Given that increased ECM synthesis was observed in low cell density microtissues, we next explored the effect of growth factors—GDF5, GDF6 in isolation or in combination with TGFβ3. A synergistic effect on DNA content (indicative of increased cell number), total sGAG and total collagen was observed for both cell types with higher values observed in groups stimulated with the addition of TGFβ3 to GDF5 or GDF6, and this was evident for both cell types. NP microtissues stimulated with TGFβ3 + GDF5 or TGFβ3 + GDF6 or TGFβ3 + GDF5 + GDF6 were comparable to TGFβ3 alone while the NC microtissues cultured in the presence of GDF5 alone were equally effective to TGFβ3 and had a synergistic effect when combined (TGFβ3 + GDF5). When these effects were evaluated on a per cell basis, the effects were diminished, with only a marginal increase observed in microtissues that were co-stimulated with growth factors. This was evident in the NC microtissues, but was not explicit in NP microtissues. Interestingly, Coleman et al. showed that MSCs co-stimulated with TGFβ3 (10 ng/mL) and GDF5 (150 ng/mL) resulted in a significant increase in sGAG/DNA by day 7 compared to TGFβ3 alone [58]. In this study, microtissues cultured in the combination of those growth factors did not result in any significant increase in proteoglycan synthesis on a per cell basis compared to TGFβ3 alone, irrespective of the cell type. This could be due to the lower concentration of GDF5 used (100 ng/mL) or could be attributed to the difference in the cell type or culturing conditions with lower oxygen and glucose concentrations. Another study by Jenner et al. showed that GDF5 stimulation caused an increase in cell numbers but collagen/DNA was significantly high in MSC scaffolds treated with TGFβ1 compared to GDF5 [59]. This trend was also observed in this study for NC microtissues, whereby TGFβ3 stimulation resulted in a significant increase in collagen/DNA compared to GDF5 only.

This study showed that GDF6 alone had little effect on DNA content (an indicator of cell number) for both NP and NC microtissues. This is in agreement with Bobacz et al., who reported no increase in cell numbers when articular chondrocytes were cultured in GDF6-supplemented media [60]. Gulati et al. showed that GDF6 at a concentration of 400 ng/mL resulted in a significant increase in proteoglycan accumulation in NP cells compared to the non-treated controls [27]. Similar observations were made in this study with both NP and NC microtissues exhibiting significantly higher sGAG/DNA in the presence of GDF6 compared to the non-treated controls.

sGAG/collagen was quantified to estimate the potential of the microtissues stimulated with various growth factors to produce the appropriate matrix type normally found in the disc. A higher ratio of sGAG composition of the ECM in comparison to the collagenous matrix is a predominant characteristic of NP tissue [45]. While no significant differences were observed for NP microtissues irrespective of the growth factor used, supplementation of NC microtissues with GDF5 or TGFβ3 alone or with the combination of growth factors (TGFβ3 + GDF5 + GDF6) yielded a significantly higher concentration. This implies that GDF5 or TGFβ3 alone could produce similar effects to the combination of growth factors indicating their potential to induce NC microtissues to form an NP-like matrix.

Recent research has shown that a combination of TGFβ1 and GDF5 significantly enhanced glycosaminoglycan content of human-derived MSCs, suggesting that this combination is optimal for MSC to NP cell induction [61]. Clarke et al. previously investigated the effect of TGFβ3, GDF5 and GDF6 growth factors on the discogenic differentiation of bone marrow- and adipose-derived MSCs (AD-MSCs). After 14 days, they observed that GDF6 resulted in increased expression of NP phenotypic genes KRT8, 18, and 19, FOX1 and CAXII compared to either TGFβ3 or GDF5 alone in both cell types with greater effects observed for AD-MSCs. However, they did not evaluate the effects of combining the growth factors [28].

In this study, GDF6 resulted in a higher sGAG/collagen ratio in NP microtissues, although this was not found to be statistically significant. On the other hand, GDF5 stimulation resulted in the highest sGAG/collagen ratio in NC microtissues, which was significantly higher than GDF6, TGFβ3 + GDF5 and TGFβ3 + GDF6. Interestingly, TGFβ3-only stimulation had a comparable effect to combinations with GDF5/GDF6. For NP microtissues, TGFβ3 also had a similar effect to other growth factors used in isolation and in combination, thereby demonstrating the potency of TGFβ3 in promoting matrix synthesis. The differential effects induced by the different growth factors, despite all belonging to the TGFβ superfamily, could be due to the difference in the signalling pathways activated. TGFβ3 is recognised by receptor type II, which activates the SMAD 2/3 signalling pathway, while GDF5 and GDF6 utilise the BMP receptor II, which activates the SMAD 1/5/8 pathway [28,62]. Activation of these distinct pathways ultimately results in different downstream signals, which may explain the differential effects observed in this study.

There are several limitations associated with the present study worth highlighting. First, longer term evaluation beyond 7 days may result in larger differences in microtissue maturation. However, this study is valuable as it demonstrates that microtissues stimulated with growth factors can produce a significant amount of ECM in a short time frame of 7 days, which would be compatible with a priming or conditioning strategy prior to implantation in the challenging disc microenvironment. We have previously shown that bone marrow stem cells (BMSCs) respond to growth factor stimulation (TGFβ3) under acidic conditions typically found in the degenerated disc (pH 6.8) [63]. A growth factor supplementation strategy with any of the growth factors assessed in this work may also prove to be beneficial. In addition, we have previously demonstrated that priming of BMSCs, NP and chondrocyte tissues with growth factors prior to a challenge or insult, confers better protection against acidic conditions, possibly due to the ECM being produced during the priming phase, thereby providing a protective niche [31,40,64]. In addition, further investigations on the gene expression profile of NCs would be valuable to identify whether typical NP markers are up-regulated.

5. Conclusions

In this study, we have developed and evaluated self-assembled microtissues stimulated with various growth factors and combinations. In general, while NP cells did respond to growth factor stimulation, NP cells were found to be relatively insensitive to the different growth factor types examined in isolation or in combination in this experimental setting. In contrast, NCs, which are more easily isolated in a less invasive manner and are more easily expandable, may prove to be an alternative cell type for NP repair and these cells could be primed with either GDF5 or TGFβ3 to enhance matrix synthesis.

Author Contributions: S.S. provided substantial contribution to study design, data acquisition data analysis and presentation, interpretation of data, drafting of the article, revising it critically and final approval. E.E.M. developed and performed the computational modelling, analysis, interpretation and presentation. C.T.B. is the overall project funding holder, takes responsibility for the integrity of the work as a whole from inception to finalised article, provided substantial contributions to study design, data presentation, interpretation of data, drafting of the article, revising it critically, and final approval. All authors have read and agreed to the published version of the manuscript.

Funding: This work was supported by a Science Foundation Ireland Career Development Award (15/CDA/3476).

Conflicts of Interest: The authors declare no conflict of interest.

References

1. Sakai, D.; Mochida, J.; Yamamoto, Y.; Nomura, T.; Okuma, M.; Nishimura, K.; Nakai, T.; Ando, K.; Hotta, T. Transplantation of mesenchymal stem cells embedded in Atelocollagen gel to the intervertebral disc: A potential therapeutic model for disc degeneration. *Biomaterials* **2003**, *24*, 3531–3541. [CrossRef]
2. Zhang, Y.; Drapeau, S.; Howard, S.A.; Thonar, E.J.; Anderson, D.G. Transplantation of goat bone marrow stromal cells to the degenerating intervertebral disc in a goat disc injury model. *Spine* **2011**, *36*, 372–377. [CrossRef]
3. Acosta, F.L., Jr.; Metz, L.; Adkisson, H.D.; Liu, J.; Carruthers-Liebenberg, E.; Milliman, C.; Maloney, M.; Lotz, J.C. Porcine intervertebral disc repair using allogeneic juvenile articular chondrocytes or mesenchymal stem cells. *Tissue Eng. Part A* **2011**, *17*, 3045–3055. [CrossRef] [PubMed]
4. Lyu, F.J.; Cheung, K.M.; Zheng, Z.; Wang, H.; Sakai, D.; Leung, V.Y. IVD progenitor cells: A new horizon for understanding disc homeostasis and repair. *Nat. Rev. Rheumatol.* **2019**, *15*, 102–112. [CrossRef] [PubMed]
5. Meisel, H.J.; Siodla, V.; Ganey, T.; Minkus, Y.; Hutton, W.C.; Alasevic, O.J. Clinical experience in cell-based therapeutics: Disc chondrocyte transplantation A treatment for degenerated or damaged intervertebral disc. *Biomol. Eng.* **2007**, *24*, 5–21. [CrossRef]
6. Smith, L.J.; Silverman, L.; Sakai, D.; Le Maitre, C.L.; Mauck, R.L.; Malhotra, N.R.; Lotz, J.C.; Buckley, C.T. Advancing cell therapies for intervertebral disc regeneration from the lab to the clinic: Recommendations of the ORS spine section. *JOR Spine* **2018**, *1*, e1036. [CrossRef]
7. Mochida, J.; Sakai, D.; Nakamura, Y.; Watanabe, T.; Yamamoto, Y.; Kato, S. Intervertebral disc repair with activated nucleus pulposus cell transplantation: A three-year, prospective clinical study of its safety. *Eur. Cells Mater.* **2015**, *29*, 202–212. [CrossRef]
8. Sakai, D.; Schol, J. Cell therapy for intervertebral disc repair: Clinical perspective. *J. Orthop. Transl.* **2017**, *9*, 8–18. [CrossRef]
9. Benneker, L.M.; Andersson, G.; Iatridis, J.C.; Sakai, D.; Hartl, R.; Ito, K.; Grad, S. Cell therapy for intervertebral disc repair: Advancing cell therapy from bench to clinics. *Eur. Cells Mater.* **2014**, *27*, 5–11. [CrossRef]
10. Vadala, G.; Sowa, G.; Hubert, M.; Gilbertson, L.G.; Denaro, V.; Kang, J.D. Mesenchymal stem cells injection in degenerated intervertebral disc: Cell leakage may induce osteophyte formation. *J. Tissue Eng. Regen. Med.* **2012**, *6*, 348–355. [CrossRef]
11. Vedicherla, S.; Buckley, C.T. In vitro extracellular matrix accumulation of nasal and articular chondrocytes for intervertebral disc repair. *Tissue Cell* **2017**, *49*, 503–513. [CrossRef] [PubMed]
12. Gay, M.H.; Mehrkens, A.; Rittmann, M.; Haug, M.; Barbero, A.; Martin, I.; Schaeren, S. Nose to back: Compatibility of nasal chondrocytes with environmental conditions mimicking a degenerated intervertebral disc. *Eur. Cells Mater.* **2019**, *37*, 214–232. [CrossRef] [PubMed]
13. Borrelli, C.; Buckley, C.T. Synergistic Effects of Acidic pH and Pro-Inflammatory Cytokines IL-1 beta and TNF-alpha for Cell-Based Intervertebral Disc Regeneration. *Appl. Sci.* **2020**, *10*, 9009. [CrossRef]
14. Rotter, N.; Bonassar, L.J.; Tobias, G.; Lebl, M.; Roy, A.K.; Vacanti, C.A. Age dependence of cellular properties of human septal cartilage: Implications for tissue engineering. *Arch. Otolaryngol. Head Neck Surg.* **2001**, *127*, 1248–1252. [CrossRef]
15. Rotter, N.; Bonassar, L.J.; Tobias, G.; Lebl, M.; Roy, A.K.; Vacanti, C.A. Age dependence of biochemical and biomechanical properties of tissue-engineered human septal cartilage. *Biomaterials* **2002**, *23*, 3087–3094. [CrossRef]
16. Pratsinis, H.; Kletsas, D. Growth factors in intervertebral disc homeostasis. *Connect. Tissue Res.* **2008**, *49*, 273–276. [CrossRef]
17. Chen, S.; Liu, S.; Ma, K.; Zhao, L.; Lin, H.; Shao, Z. TGF-beta signaling in intervertebral disc health and disease. *Osteoarthr. Cartil.* **2019**, *27*, 1109–1117. [CrossRef]
18. Bian, Q.; Ma, L.; Jain, A.; Crane, J.L.; Kebaish, K.; Wan, M.; Zhang, Z.; Edward Guo, X.; Sponseller, P.D.; Seguin, C.A.; et al. Mechanosignaling activation of TGFbeta maintains intervertebral disc homeostasis. *Bone Res.* **2017**, *5*, 17008. [CrossRef]
19. Risbud, M.V.; Di Martino, A.; Guttapalli, A.; Seghatoleslami, R.; Denaro, V.; Vaccaro, A.R.; Albert, T.J.; Shapiro, I.M. Toward an optimum system for intervertebral disc organ culture: TGF-beta 3 enhances nucleus pulposus and anulus fibrosus survival and function through modulation of TGF-beta-R expression and ERK signaling. *Spine* **2006**, *31*, 884–890. [CrossRef]
20. Kim, M.J.; Lee, J.H.; Kim, J.S.; Kim, H.Y.; Lee, H.C.; Byun, J.H.; Lee, J.H.; Kim, N.H.; Oh, S.H. Intervertebral Disc Regeneration Using Stem Cell/Growth Factor-Loaded Porous Particles with a Leaf-Stacked Structure. *Biomacromolecules* **2020**, *21*, 4795–4805. [CrossRef]
21. Le Maitre, C.L.; Freemont, A.J.; Hoyland, J.A. Localization of degradative enzymes and their inhibitors in the degenerate human intervertebral disc. *J. Pathol.* **2004**, *204*, 47–54. [CrossRef] [PubMed]
22. Settle, S.H., Jr.; Rountree, R.B.; Sinha, A.; Thacker, A.; Higgins, K.; Kingsley, D.M. Multiple joint and skeletal patterning defects caused by single and double mutations in the mouse Gdf6 and Gdf5 genes. *Dev. Biol.* **2003**, *254*, 116–130. [CrossRef]

23. Wolfman, N.M.; Hattersley, G.; Cox, K.; Celeste, A.J.; Nelson, R.; Yamaji, N.; Dube, J.L.; DiBlasio-Smith, E.; Nove, J.; Song, J.J.; et al. Ectopic induction of tendon and ligament in rats by growth and differentiation factors 5, 6, and 7, members of the TGF-beta gene family. *J. Clin. Investig.* **1997**, *100*, 321–330. [CrossRef]
24. Enochson, L.; Stenberg, J.; Brittberg, M.; Lindahl, A. GDF5 reduces MMP13 expression in human chondrocytes via DKK1 mediated canonical Wnt signaling inhibition. *Osteoarthr. Cartil.* **2014**, *22*, 566–577. [CrossRef] [PubMed]
25. Guo, S.; Cui, L.; Xiao, C.; Wang, C.; Zhu, B.; Liu, X.; Li, Y.; Liu, X.; Wang, D.; Li, S. The Mechanisms and Functions of GDF-5 in Intervertebral Disc Degeneration. *Orthop. Surg.* **2021**, *13*, 734–741. [CrossRef] [PubMed]
26. Risbud, M.V.; Shapiro, I.M. Role of cytokines in intervertebral disc degeneration: Pain and disc content. *Nat. Rev. Rheumatol.* **2014**, *10*, 44–56. [CrossRef] [PubMed]
27. Gulati, T.; Chung, S.A.; Wei, A.Q.; Diwan, A.D. Localization of bone morphogenetic protein 13 in human intervertebral disc and its molecular and functional effects in vitro in 3D culture. *J. Orthop. Res.* **2015**, *33*, 1769–1775. [CrossRef] [PubMed]
28. Clarke, L.E.; McConnell, J.C.; Sherratt, M.J.; Derby, B.; Richardson, S.M.; Hoyland, J.A. Growth differentiation factor 6 and transforming growth factor-beta differentially mediate mesenchymal stem cell differentiation, composition, and micromechanical properties of nucleus pulposus constructs. *Arthritis Res.* **2014**, *16*, R67. [CrossRef] [PubMed]
29. Vedicherla, S.; Buckley, C.T. Rapid Chondrocyte Isolation for Tissue Engineering Applications: The Effect of Enzyme Concentration and Temporal Exposure on the Matrix Forming Capacity of Nasal Derived Chondrocytes. *BioMed Res. Int.* **2017**, *2017*, 2395138. [CrossRef] [PubMed]
30. Futrega, K.; Palmer, J.S.; Kinney, M.; Lott, W.B.; Ungrin, M.D.; Zandstra, P.W.; Doran, M.R. The microwell-mesh: A novel device and protocol for the high throughput manufacturing of cartilage microtissues. *Biomaterials* **2015**, *62*, 1–12. [CrossRef] [PubMed]
31. Naqvi, S.M.; Gansau, J.; Buckley, C.T. Priming and cryopreservation of microencapsulated marrow stromal cells as a strategy for intervertebral disc regeneration. *Biomed. Mater.* **2018**, *13*, 034106. [CrossRef] [PubMed]
32. McDonnell, E.E.; Buckley, C.T. Investigating the physiological relevance of ex vivo disc organ culture nutrient microenvironments using in silico modeling and experimental validation. *JOR Spine* **2021**, *4*, e1141. [CrossRef] [PubMed]
33. Bibby, S.R.; Jones, D.A.; Ripley, R.M.; Urban, J.P. Metabolism of the intervertebral disc: Effects of low levels of oxygen, glucose, and pH on rates of energy metabolism of bovine nucleus pulposus cells. *Spine* **2005**, *30*, 487–496. [CrossRef] [PubMed]
34. Huang, C.Y.; Yuan, T.Y.; Jackson, A.R.; Hazbun, L.; Fraker, C.; Gu, W.Y. Effects of low glucose concentrations on oxygen consumption rates of intervertebral disc cells. *Spine* **2007**, *32*, 2063–2069. [CrossRef]
35. Huang, C.Y.; Gu, W.Y. Effects of mechanical compression on metabolism and distribution of oxygen and lactate in intervertebral disc. *J Biomech* **2008**, *41*, 1184–1196. [CrossRef]
36. Kafienah, W.; Sims, T.J. Biochemical methods for the analysis of tissue-engineered cartilage. *Methods Mol. Biol.* **2004**, *238*, 217–230.
37. Gupta, M.S.; Cooper, E.S.; Nicoll, S.B. Transforming growth factor-beta 3 stimulates cartilage matrix elaboration by human marrow-derived stromal cells encapsulated in photocrosslinked carboxymethylcellulose hydrogels: Potential for nucleus pulposus replacement. *Tissue Eng. Part A* **2011**, *17*, 2903–2910. [CrossRef]
38. Pelle, D.W.; Peacock, J.D.; Schmidt, C.L.; Kampfschulte, K.; Scholten, D.J., 2nd; Russo, S.S.; Easton, K.J.; Steensma, M.R. Genetic and functional studies of the intervertebral disc: A novel murine intervertebral disc model. *PLoS ONE* **2014**, *9*, e112454. [CrossRef]
39. Naqvi, S.M.; Buckley, C.T. Differential response of encapsulated nucleus pulposus and bone marrow stem cells in isolation and coculture in alginate and chitosan hydrogels. *Tissue Eng. Part A* **2015**, *21*, 288–299. [CrossRef]
40. Gansau, J.; Buckley, C.T. Priming as a strategy to overcome detrimental pH effects on cells for intervertebral disc regeneration. *Eur. Cells Mater.* **2021**, *41*, 153–169. [CrossRef]
41. Coleman, C.M.; Tuan, R.S. Functional role of growth/differentiation factor 5 in chondrogenesis of limb mesenchymal cells. *Mech. Dev.* **2003**, *120*, 823–836. [CrossRef]
42. Nishitoh, H.; Ichijo, H.; Kimura, M.; Matsumoto, T.; Makishima, F.; Yamaguchi, A.; Yamashita, H.; Enomoto, S.; Miyazono, K. Identification of type I and type II serine/threonine kinase receptors for growth/differentiation factor-5. *J. Biol. Chem.* **1996**, *271*, 21345–21352. [CrossRef] [PubMed]
43. Zou, H.; Wieser, R.; Massague, J.; Niswander, L. Distinct roles of type I bone morphogenetic protein receptors in the formation and differentiation of cartilage. *Genes Dev.* **1997**, *11*, 2191–2203. [CrossRef] [PubMed]
44. Gooch, K.J.; Blunk, T.; Courter, D.L.; Sieminski, A.L.; Vunjak-Novakovic, G.; Freed, L.E. Bone morphogenetic proteins-2, -12, and -13 modulate in vitro development of engineered cartilage. *Tissue Eng.* **2002**, *8*, 591–601. [CrossRef] [PubMed]
45. Hodgkinson, T.; Shen, B.; Diwan, A.; Hoyland, J.A.; Richardson, S.M. Therapeutic potential of growth differentiation factors in the treatment of degenerative disc diseases. *JOR Spine* **2019**, *2*, e1045. [CrossRef] [PubMed]
46. Risbud, M.V.; Schoepflin, Z.R.; Mwale, F.; Kandel, R.A.; Grad, S.; Iatridis, J.C.; Sakai, D.; Hoyland, J.A. Defining the phenotype of young healthy nucleus pulposus cells: Recommendations of the Spine Research Interest Group at the 2014 annual ORS meeting. *J. Orthop. Res.* **2015**, *33*, 283–293. [CrossRef]
47. Wen-tao, Q.; Ying, Z.; Juan, M.; Xin, G.; Yu-bing, X.; Wei, W.; Xiaojun, M. Optimization of the cell seeding density and modeling of cell growth and metabolism using the modified Gompertz model for microencapsulated animal cell culture. *Biotechnol. Bioeng.* **2006**, *93*, 887–895. [CrossRef]
48. Foldager, C.B.; Gomoll, A.H.; Lind, M.; Spector, M. Cell Seeding Densities in Autologous Chondrocyte Implantation Techniques for Cartilage Repair. *Cartilage* **2012**, *3*, 108–117. [CrossRef]

49. Zamani, S.; Hashemibeni, B.; Esfandiari, E.; Kabiri, A.; Rabbani, H.; Abutorabi, R. Assessment of TGF-beta3 on production of aggrecan by human articular chondrocytes in pellet culture system. *Adv. Biomed. Res.* **2014**, *3*, 54. [CrossRef]
50. Chen, J.; Wang, Y.; Chen, C.; Lian, C.; Zhou, T.; Gao, B.; Wu, Z.; Xu, C. Exogenous Heparan Sulfate Enhances the TGF-beta3-Induced Chondrogenesis in Human Mesenchymal Stem Cells by Activating TGF-beta/Smad Signaling. *Stem. Cells Int.* **2016**, *2016*, 1520136. [CrossRef]
51. Byers, B.A.; Mauck, R.L.; Chiang, I.E.; Tuan, R.S. Transient exposure to transforming growth factor beta 3 under serum-free conditions enhances the biomechanical and biochemical maturation of tissue-engineered cartilage. *Tissue Eng. Part A* **2008**, *14*, 1821–1834. [CrossRef]
52. Huang, A.H.; Stein, A.; Tuan, R.S.; Mauck, R.L. Transient exposure to transforming growth factor beta 3 improves the mechanical properties of mesenchymal stem cell-laden cartilage constructs in a density-dependent manner. *Tissue Eng. Part A* **2009**, *15*, 3461–3472. [CrossRef]
53. Afizah, H.; Yang, Z.; Hui, J.H.; Ouyang, H.W.; Lee, E.H. A comparison between the chondrogenic potential of human bone marrow stem cells (BMSCs) and adipose-derived stem cells (ADSCs) taken from the same donors. *Tissue Eng.* **2007**, *13*, 659–666. [CrossRef]
54. Amin, H.D.; Brady, M.A.; St-Pierre, J.P.; Stevens, M.M.; Overby, D.R.; Ethier, C.R. Stimulation of chondrogenic differentiation of adult human bone marrow-derived stromal cells by a moderate-strength static magnetic field. *Tissue Eng. Part A* **2014**, *20*, 1612–1620. [CrossRef]
55. Li, S.; Macon, A.L.B.; Jacquemin, M.; Stevens, M.M.; Jones, J.R. Sol-gel derived lithium-releasing glass for cartilage regeneration. *J. Biomater. Appl.* **2017**, *32*, 104–113. [CrossRef]
56. Li, D.; Ma, X.; Zhao, T. Mechanism of TGF-beta3 promoting chondrogenesis in human fat stem cells. *Biochem. Biophys. Res. Commun.* **2020**, *530*, 725–731. [CrossRef]
57. Furumatsu, T.; Tsuda, M.; Taniguchi, N.; Tajima, Y.; Asahara, H. Smad3 induces chondrogenesis through the activation of SOX9 via CREB-binding protein/p300 recruitment. *J. Biol. Chem.* **2005**, *280*, 8343–8350. [CrossRef]
58. Coleman, C.M.; Vaughan, E.E.; Browe, D.C.; Mooney, E.; Howard, L.; Barry, F. Growth differentiation factor-5 enhances in vitro mesenchymal stromal cell chondrogenesis and hypertrophy. *Stem. Cells Dev.* **2013**, *22*, 1968–1976. [CrossRef]
59. Jenner, J.M.; van Eijk, F.; Saris, D.B.; Willems, W.J.; Dhert, W.J.; Creemers, L.B. Effect of transforming growth factor-beta and growth differentiation factor-5 on proliferation and matrix production by human bone marrow stromal cells cultured on braided poly lactic-co-glycolic acid scaffolds for ligament tissue engineering. *Tissue Eng.* **2007**, *13*, 1573–1582. [CrossRef]
60. Bobacz, K.; Gruber, R.; Soleiman, A.; Erlacher, L.; Smolen, J.S.; Graninger, W.B. Expression of bone morphogenetic protein 6 in healthy and osteoarthritic human articular chondrocytes and stimulation of matrix synthesis in vitro. *Arthritis Rheum.* **2003**, *48*, 2501–2508. [CrossRef]
61. Morita, K.; Schol, J.; Volleman, T.N.E.; Sakai, D.; Sato, M.; Watanabe, M. Screening for Growth-Factor Combinations Enabling Synergistic Differentiation of Human MSC to Nucleus Pulposus Cell-Like Cells. *Appl. Sci.* **2021**, *11*, 3673. [CrossRef]
62. Mazerbourg, S.; Sangkuhl, K.; Luo, C.W.; Sudo, S.; Klein, C.; Hsueh, A.J. Identification of receptors and signaling pathways for orphan bone morphogenetic protein/growth differentiation factor ligands based on genomic analyses. *J. Biol. Chem.* **2005**, *280*, 32122–32132. [CrossRef]
63. Naqvi, S.M.; Buckley, C.T. Bone Marrow Stem Cells in Response to Intervertebral Disc-Like Matrix Acidity and Oxygen Concentration: Implications for Cell-based Regenerative Therapy. *Spine* **2016**, *41*, 743–750. [CrossRef]
64. Naqvi, S.M.; Gansau, J.; Gibbons, D.; Buckley, C.T. In vitro co-culture and ex vivo organ culture assessment of primed and cryopreserved stromal cell microcapsules for IVD regeneration. *Eur. Cells Mater.* **2019**, *37*, 134–152. [CrossRef]

Article

Single Cell RNA-Sequence Analyses Reveal Uniquely Expressed Genes and Heterogeneous Immune Cell Involvement in the Rat Model of Intervertebral Disc Degeneration

Milad Rohanifar [1], Sade W. Clayton [2], Garrett W.D. Easson [2], Deepanjali S. Patil [1], Frank Lee [1], Liufang Jing [1], Marcos N. Barcellona [1], Julie E. Speer [1], Jordan J. Stivers [2], Simon Y. Tang [2] and Lori A. Setton [1,2,*]

1 Department of Biomedical Engineering, Washington University in St. Louis, St. Louis, MO 63130, USA
2 Department of Orthopedic Surgery, Washington University School of Medicine, St. Louis, MO 63110, USA
* Correspondence: setton@wustl.edu

Abstract: Intervertebral disc (IVD) degeneration is characterized by a loss of cellularity, and changes in cell-mediated activity that drives anatomic changes to IVD structure. In this study, we used single-cell RNA-sequencing analysis of degenerating tissues of the rat IVD following lumbar disc puncture. Two control, uninjured IVDs (L2-3, L3-4) and two degenerated, injured IVDs (L4-5, L5-6) from each animal were examined either at the two- or eight-week post-operative time points. The cells from these IVDs were extracted and transcriptionally profiled at the single-cell resolution. Unsupervised cluster analysis revealed the presence of four known cell types in both non-degenerative and degenerated IVDs based on previously established gene markers: IVD cells, endothelial cells, myeloid cells, and lymphoid cells. As a majority of cells were associated with the IVD cell cluster, sub-clustering was used to further identify the cell populations of the nucleus pulposus, inner and outer annulus fibrosus. The most notable difference between control and degenerated IVDs was the increase of myeloid and lymphoid cells in degenerated samples at two- and eight-weeks post-surgery. Differential gene expression analysis revealed multiple distinct cell types from the myeloid and lymphoid lineages, most notably macrophages and B lymphocytes, and demonstrated a high degree of immune specificity during degeneration. In addition to the heterogenous infiltrating immune cell populations in the degenerating IVD, the increased number of cells in the AF sub-cluster expressing *Ngf* and *Ngfr*, encoding for p75NTR, suggest that NGF signaling may be one of the key mediators of the IVD crosstalk between immune and neuronal cell populations. These findings provide the basis for future work to understand the involvement of select subsets of non-resident cells in IVD degeneration.

Keywords: intervertebral disc degeneration; single-cell RNA sequencing; cell type

Citation: Rohanifar, M.; Clayton, S.W.; Easson, G.W.; Patil, D.S.; Lee, F.; Jing, L.; Barcellona, M.N.; Speer, J.E.; Stivers, J.J.; Tang, S.Y.; et al. Single Cell RNA-Sequence Analyses Reveal Uniquely Expressed Genes and Heterogeneous Immune Cell Involvement in the Rat Model of Intervertebral Disc Degeneration. *Appl. Sci.* **2022**, *12*, 8244. https://doi.org/10.3390/app12168244

Academic Editor: Benjamin Gantenbein

Received: 3 July 2022
Accepted: 18 July 2022
Published: 18 August 2022

1. Introduction

The intervertebral disc (IVD) provides for motion and flexibility in the spine [1]. Degeneration of the IVD is characterized by a loss of cellularity, changes in composition, and loss of hydration that is manifested as changes in disc height and MRI signal intensity. These features of IVD degeneration in the lumbar and cervical spine can contribute to mal-alignment of the adjoining vertebral bodies and affect the integrity and health of the IVD and adjacent neural structures, leaving the IVD vulnerable to progressive damage during the loading conditions of daily living [2–4]. The cell-mediated changes affecting IVD growth, extracellular matrix synthesis, and metabolism can affect IVD proteolytic degradation, and are of great interest for identifying the relevant cellular and molecular targets to support IVD repair and regeneration.

The cell density of the nucleus pulposus (NP) region is at the lowest in the IVD, and are the principal contributors to the loss of cellularity with age. NP cells have long been considered to share features with chondrocytes in the adult IVD, due to their roundedness and high expression of multiple chondrocyte markers including SOX9, type II collagen, and aggrecan upon bulk mRNA transcriptional profiling [5–9]. NP cells are derived from the embryonic notochord and express transcriptionally unique markers that reflect their notochordal origins, including CD24, cytokeratins, and the brachyury and FOX transcription factors [10–13]. Cells of the annulus fibrosus (AF) are derived from mesenchyme and express some overlapping and some differing molecular markers from adjacent NP cells. Indeed, there have been numerous studies designed to identify the unique cellular phenotypes of NP and AF cells, as well as non-resident cells involved in degeneration and repair, using tools of bulk RNA transcription or RNA-sequencing, proteomic profiling, flow cytometry, and protein assays [5,14–18]. The improved understanding of cell-specific molecular changes is crucial for identifying important protein targets that drive progress degeneration and molecular targets for stem cell-mediated IVD repair.

In recent years, single-cell RNA sequencing (scRNA-seq) of IVD cells, both from anatomically distinct regions and for comparisons with pathology, has been used to characterize the phenotype and abundance of distinct cell populations in the IVD in human and animal tissue sources [5,15,17,19–21]. When cells are isolated rapidly and immediately prepared for RNA-sequencing, this technique has the potential to reveal the endogenous RNA expression of resident cells and thus identify differences in cell sub-populations in the native IVD [5,15,17,19–22]. Results of sc-RNA-seq from rat and bovine IVD tissues, and from non-degenerate human IVD tissues, rely on mapping cells with similar transcriptomic profiles into "clusters" in order to name and number discrete cell populations in the non-degenerate IVD. While the number of clusters is highly variable in the absence of a consensus-based approach, it is most commonly observed that the majority of IVD cells map to a few clusters based on their expression of common genes (50–99% of isolated cells). Studies have used this approach to identify small but meaningful stem cell or progenitor cell populations in the IVD [15,20], or to identify relationships between identified clusters and primary cell functions of matrix regulation, stress responses, cell cycle, and more [22].

Only a few studies have used sc-RNA-seq to better understand the progression of IVD degeneration at the cellular level. In donor human tissues, sc-RNA-seq has been used to identify genes associated with cells of the degenerative IVD with some findings that corroborate prior work including the unique associative expression for CTGF, S100A1/A2, and TNF receptors in cells from degenerated IVDs [5,22]; both human studies and a study of a rodent bipedal IVD degeneration model also present surprising findings for novel gene involvement that have not yet been confirmed in other studies or species. In some cases, sc-RNA-seq identifies cells that express macrophage markers at high levels [5] consistent with prior literature showing increased involvement of macrophages in animal models and human tissues with IVD degeneration [23]. The IVD is an aneural, alymphatic, and avascular structure upon the termination of development and growth, however, it can remain immunologically separate from the host over the lifetime of an individual. The presence of macrophage markers in degenerated IVD tissues and cells suggests integrity of the IVD may be disrupted such that cells of the systemic circulation may infiltrate and reside in the IVD. These cells can be expected to be minor and difficult to localize via immunostaining or RT-PCR; for this reason, sc-RNA-seq provides the potential to identify the unique phenotypes of these relatively small cell populations in the IVD.

For these reasons, we performed scRNA-seq analysis on cells obtained from rat lumbar IVD tissues at two time points following induction of IVD degeneration via stab injury [24–28]. In brief, we identified four major cell types in the non-degenerate and degenerative IVD tissues for 2 weeks following surgery, based on the expression of "classical" cell-specific markers for NP and AF cells as identified previously [5]. The cell clustering scheme was held fixed between control and IVD degeneration conditions to estimate the differences in cell numbers and major gene expression levels for respective cell populations within each cluster. Further, we used sub-clustering to identify discrete native cell types within the IVD, and gene set analysis to identify the numerically minor cells of the degenerated IVD and their changes from 2 to 8 weeks after induced IVD degeneration. This study is innovative for utilizing sc-RNA-seq to reveal both temporal and spatial changes to cell presence in the IVD following onset of disc degeneration, and for presenting data on a specimen-specific basis.

2. Materials and Methods

2.1. IVD Tissue Preparation and Single-Cell Isolation

Rats (n = 8, male Sprague-Dawley, 8-weeks-old) underwent surgery for retroperitoneal exposure of the L4-L6 lumbar spine in protocols approved by the Washington University IACUC. The L4-5 and L5-6 IVDs each received a single unilateral stab injury via a 27G needle. The rats were allowed to recover postoperatively for either 2 (n = 4) or 8 weeks (n = 4) (Figure 1). Following the respective recovery periods, the animals were euthanized and the IVD tissues were isolated and pooled from L4-5 and L5-6 (LDP = Lumbar disc puncture), or from L2-3 and L3-4 (CON = control), with care to remove peripheral muscle, ligaments, and attachments by microdissection. The isolated IVD tissue was digested in medium containing 0.2% type 2 collagenase (Worthington Biochemical, Lakewood, NJ, USA) and 0.3% pronase (Roche, Basel, Switzerland) for <4 h total at 37 °C and 5% CO_2. After digestion, a cell pellet was obtained by centrifuging the medium for 10 min at 400 rcf. The media was aspirated, and cells were resuspended in PBS (Sigma Aldrich, St. Louis, MO, USA), followed by filtering through a 70 µm filter to get rid of undesired cell debris. The flowthrough was again centrifuged for 10 min at 400 rcf, and the resulting cell pellet was resuspended in PBS. The cells isolated from pooled IVDs for each rat were kept together and on ice and considered as a separate sample for a total of n = 16 samples in total (n = 4 for CON and LDP at 2 weeks post-surgery; n = 4 for CON and LDP at 8 weeks post-surgery). Samples of IVD cells were immediately transported on ice to the sequencing facility. More than 80% of cells were determined to be viable. All analyses described below were performed on individual rat cell samples unless otherwise indicated.

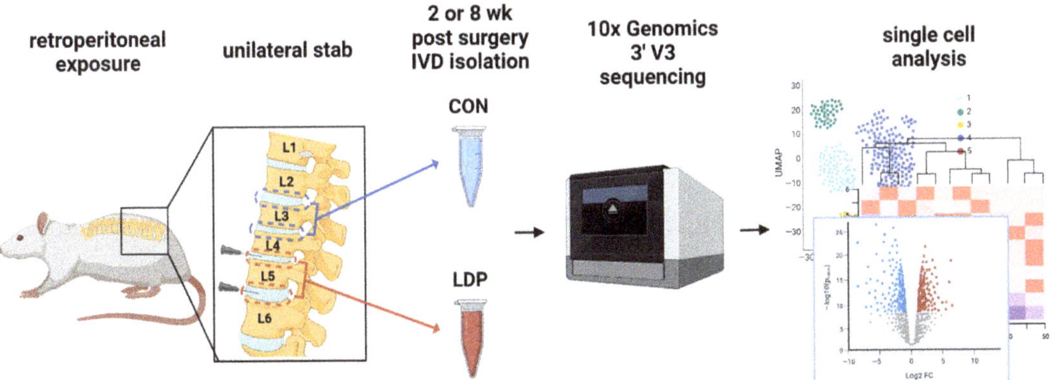

Figure 1. Flowchart depicting an overview of the single-cell RNA-seq experimental design.

2.2. cDNA Library Generation and Single-Cell RNA-Sequencing (scRNA-Seq) with Data Standardization

The input samples were submitted to the Washington University Genome Technology Access Center to obtain and sequence the cDNA libraries (10xGenomics, 3'v3.1; Illumina NovaSeq S4) according to established protocols. For CON and LDP samples, only cells with <10% mitochondrial features, gene counts of 500–3000 and UMI (unique molecular identifier) counts between 500–30,000 were determined to be of sufficient quality to include in the analyses (Partek Flow™, Partek Inc., St. Louis, MO, USA). Under these selection criteria, the total number of cells from rat samples was 42,560 and 51,365 for CON and LDP at 2 weeks post-surgery, 27,064 and 28,293 for CON and LDP at 8 weeks post-surgery, respectively.

2.3. Dimensionality Reduction and Unsupervised Clustering

In order to identify resident cell populations, we first performed Principal Components Analysis (PCA) on pooled data from n = 4 CON samples at 2 weeks, and then again for the 8 weeks samples. PCA was performed to reduce dimensionality of the data for CON at each timepoint with selection of optimum number of principal components after normalization; this process was repeated for the n = 4 LDP samples at 2 and at 8 weeks. Cluster analysis was then performed with the iterative procedure of K-means clustering on CON cells at 2 weeks to identify optimal separation of cell clusters. A total of five major cell clusters were identified as providing for optimal separation of the data for CON samples. The percentage of cells mapped to each cluster was evaluated for each tissue sample at both 2 and 8 weeks post-surgery as a test of inter-subject variability and appropriateness of the clustering scheme. Annotated cluster data were visualized with uniform manifold approximation and projection (UMAP) analyses.

2.4. Cell Identification

The nomenclature classifying cell types within each identified cluster for CON samples at 2 weeks was based on a set of "marker genes" based on reports within the literature (see Table 1). For CON samples, we first validated the presence of known cell subsets based on prior scRNA-seq studies of intervertebral cells, endothelial cells, myeloid and lymphoid cells of the rat IVD [5,15]. Violin plots of gene expression levels for all cells within each cluster were used to define the majority cell type in each cluster.

Table 1. Genes examined for definition of cell subtypes in the IVD of rat samples (i.e., marker genes). Percentages are numbers of total cells assigned to each cluster averaged across four rats. CON—cells from control (unoperated) IVDs; LDP—cells from lumbar disc punctured IVDs at 2 or 8 weeks following surgery.

Cell Type	CON	LDP	Gene Marker
Intervertebral disc cells	2 week (87.7%) 8 week (83.7%)	2 week (82.4%) 8 week (87.4%)	*Can*—aggrecan *Col2a1*—collagen type II alpha 1 chain *Sox9*—SRY-box 9 *Fmod*—fibromodulin *Comp*—cartilage oligomeric matrix protein
Endothelial cells	2 week (7.3%) 8 week (0%)	2 week (6.41%) 8 week (2.3%)	*Cdh5*—cadherin 5 *Plvap*—plasmalemma vesicle-associated protein *Ifitm3*—interferon-induced transmembrane protein 3 *Flt1*—tyrosine-protein kinase receptor
Myeloid cells	2 week (1.0%) 8 week (1.2%)	2 week (1.4%) 8 week (0.8%)	*Cd14*—monocyte differentiation antigen CD14 *Mrc1*—mannose receptor C-Type 1 *Cd68*—CD68 antigen *Lyz2*—lysozyme 2
Lymphoid cells	2 week (0.2%) 8 week (8.4%)	2 week (8.2%) 8 week (5.1%)	*Cd79b*—B-cell antigen receptor complex-associated protein beta chain *Blnk*—B cell linker *Ptprc*—protein tyrosine phosphatase receptor type C *Myb*—MYB proto-oncogene, transcription factor

As a majority of cells in the CON populations were identified with a cluster named intervertebral disc cells, we sought to further sub-cluster this population in a post-hoc process separate from the data for cells mapped to the other clusters. IVD cells for CON samples at 2 weeks were subjected to PCA to select an optimal number of principal components, followed by K-means clustering on pooled data to identify further separation of cells in the intervertebral disc clusters based on the lowest Davies–Bouldin index.

The above-described process for clustering and sub-clustering was repeated for CON samples at 8 weeks post-surgery following the same cluster nomenclature and marker genes. We sought to retain five major cell clusters for samples from the LDP population at both 2 and 8 weeks post-surgery, in order to directly compare cell numbers and gene expression levels against those of the CON population. In large part, the nomenclature for these five cell clusters did not differ from that of CON populations; however, the numbers of cells mapped to each cluster and gene expression for marker and novel genes varied substantially.

2.5. Gene Specific Analysis

Differential gene expression analyses were performed with the GSA toolbox in Partek Flow^{TM} to identify the differentially expressed genes within each rat sample between CON and LDP at each of 2 and 8 weeks post-surgery. The genes exhibiting the greatest fold-change were identified ($\log_2$ (fold change)) for each sample at 2 weeks or 8 weeks with a statistical significance of false discovery rate (FDR) p-value < 0.001. This set of genes identified as most highly differentially expressed differed among samples with some overlap as described in the results. Genes with the highest fold-difference between CON and LDP that are common across samples at 2 or 8 weeks are presented here.

2.6. Gene Set Analyses

Gene set analyses were performed using the Gene Ontology resource (https://www.uniprot.org/help/gene_ontology (accessed on and after 1 September 2021) (UniProt Database) to predict the physiological roles and molecular processes of the upregulated gene networks. The top 50 differentially regulated genes for the respective cluster or subclusters were used to define the relevant cellular processes.

2.7. Statistical Analyses

Optimal separation of cells into a minimum number of distinct clusters was identified via the minimal Davies-Bouldin index, as indicated above for CON cells at 2 weeks and for IVD cells when separated into sub-clusters. Data for the magnitude of gene expression for cell-specific markers, e.g., genes defining B and T lymphocytes, were tested for differences between CON and LDP populations at each of 2 and 8 weeks post-surgery by the Student's t-test or the Mann–Whitney test as appropriate (GraphPad Prism^{TM}). Likewise, similar comparisons were made for the expression of *Ngf* and *Ngfr* (p75NTR) in matched sub-clusters between CON and LDP populations. Chi-squared tests were used to compare the proportions between matched cell clusters/subclusters across CON and LDP populations. Gene set enrichment analyses were performed by matching genes to defined pathways based on known cell functions and reporting overrepresented cell numbers in groups via a Fisher's exact test Partek Flow^{TM}.

3. Results

3.1. Cell Populations Identified in IVD Tissues from 2 Weeks Post-Surgery

Samples in control IVD tissues 2 weeks post-surgery. Unsupervised K-means clustering was used for the CON samples to group the cells into clusters with equal variance. Results show that five clusters separated all cells into discrete populations. There were identified four named clusters: intervertebral disc cells, endothelial cells, myeloid, and lymphoid cells (Figure 2a); cluster 5 was annotated as "other cells" due to the lack of common identification markers. Identification of cell populations within each cluster was based on known gene markers that were pre-selected as described in methods (Figure 2b). The identified clusters had a distribution of cell counts that was very similar across samples for the four CON samples derived from four separate rats (Figure 2c). Each sample showed a similar cell distribution profile where intervertebral disc cells contributed to most of the cluster (~87.7%), followed by endothelial cells (~7.3%), myeloid (~1.0%) and lymphoid cells (~0.2%).

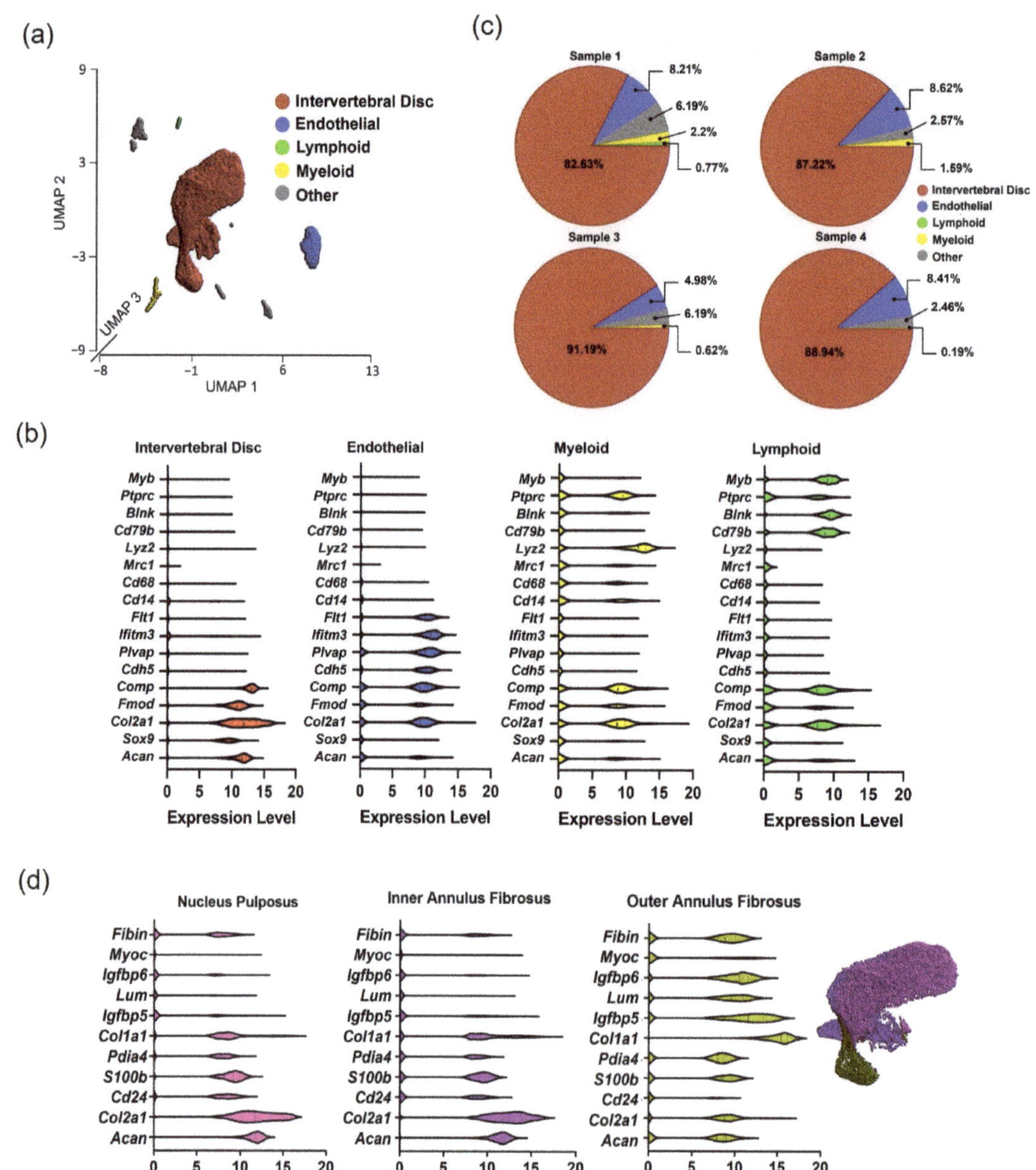

Figure 2. (**a**) UMAP plot representation of five cell types within CON group at two weeks after surgery. (**b**) Expression levels of marker genes within intervertebral disc cells, endothelial cells, myeloid, and lymphoid cells. (**c**) Pie chart showing the distribution of each cell analyzed for the four rat samples in the CON group. (**d**) The expression level of marker genes for three subclusters of intervertebral disc cells: nucleus pulposus, inner annulus fibrosus, and outer annulus fibrosus.

The selected marker genes were chosen based on *a priori* knowledge for identifying the discrete cell clusters (Figure 2b). There was some commonality in extracellular matrix genes across clusters, primarily for *Col2a1* and *Comp*; this observation is consistent with the prevalent expression of these extracellular matrix proteins in different tissues and suggests that relying on non-extracellular matrix genes as differentiating markers may be preferable. Markers selected as lymphoid and myeloid markers proved to be effective in separating cells into clusters with more separation.

As more than 85% of all cells mapped to the "intervertebral disc cells" cluster, we sought to sub-cluster this population to identify the known cell populations. The IVD cell cluster was partitioned into three distinctively adjacent subclusters that were classified as nucleus pulposus (NP), inner annulus fibrosus (IAF), and outer annulus fibrosus (OAF) based on relative expression levels of classical IVD cell markers (Figure 2d). Subcluster 1 was identified to be NP cells based on a relatively higher expression for *Acan* and *Col2a1* while also expressing lower *Col1a1* compared to other IVD cells [8,29–32]. Additionally, these cells highly express *Cd24*, a classical marker for NP cells [13,33–35], previously identified markers of the chondrocyte lineage including S100B, and a novel marker for bovine NP tissues, PDIA4 [19,36,37]. Cells mapping to the IAF cluster were similar in the pattern of expression to that of NP cells, differing with modestly increased *Col1a1* expression and decreased presence of *Cd24*. Finally, OAF cells were identified based on the expression of a number of previously identified markers for this cell type including *Igfbp5*, *Igfbp6*, *Lum*, *Myoc*, and *Fibin* (Figure 2d) [8,19,29,31,32,38].

Samples in degenerated IVD tissues. Samples harvested from the intervertebral discs of the LDP rats at 2 weeks post-surgery underwent unsupervised K-means clustering as applied to the CON group. Five clusters provided for optimal separation of the LDP cells based on known gene markers (Figure 3a,b). The IVD cell cluster contributed to the greatest proportion of the cells (~82.4%), followed by endothelial (~6.4%), myeloid (1.4%), and lymphoid clusters (~8.2%). The myeloid and lymphoid clusters exhibited the most notable difference 2 weeks after LDP (Figure 3c).

Differential gene expression between LDP and CON samples on a subject-specific basis. To directly assess differences in mRNA levels for cells from the CON (L2-3 and L3-4 levels) and LDP (L4-5 and L5-6 levels) groups within each rat sample, we performed specimen-specific differential gene expression (Figure 4a,b). Differential gene analyses demonstrated that between 700–900 genes were more highly expressed in LDP cells at 2 weeks after surgery, as compared to their counterparts amongst the CON cell population (FDR p-value < 0.001, Figure 4c). Analyses also showed between 250 and 1200 genes to be more highly expressed in CON cells as compared to the LDP cell population at 2 weeks after surgery (FDR p-value < 0.001, Figure 4c). While there was variability in differentially expressed genes across samples, we identified a subset of 25 commonly upregulated genes in LDP populations of cells from two samples with the highest total numbers of sequenced cells (i.e., samples #3 and #4); expression levels for these genes against their cell identification are shown in the heat map in Figure 4d (representative results shown for sample #4). Of these 25 genes with upregulated expression levels in LDP compared to CON cells, the majority (22/25) were associated with cells of the lymphoid or myeloid cell clusters (Figure 4d).

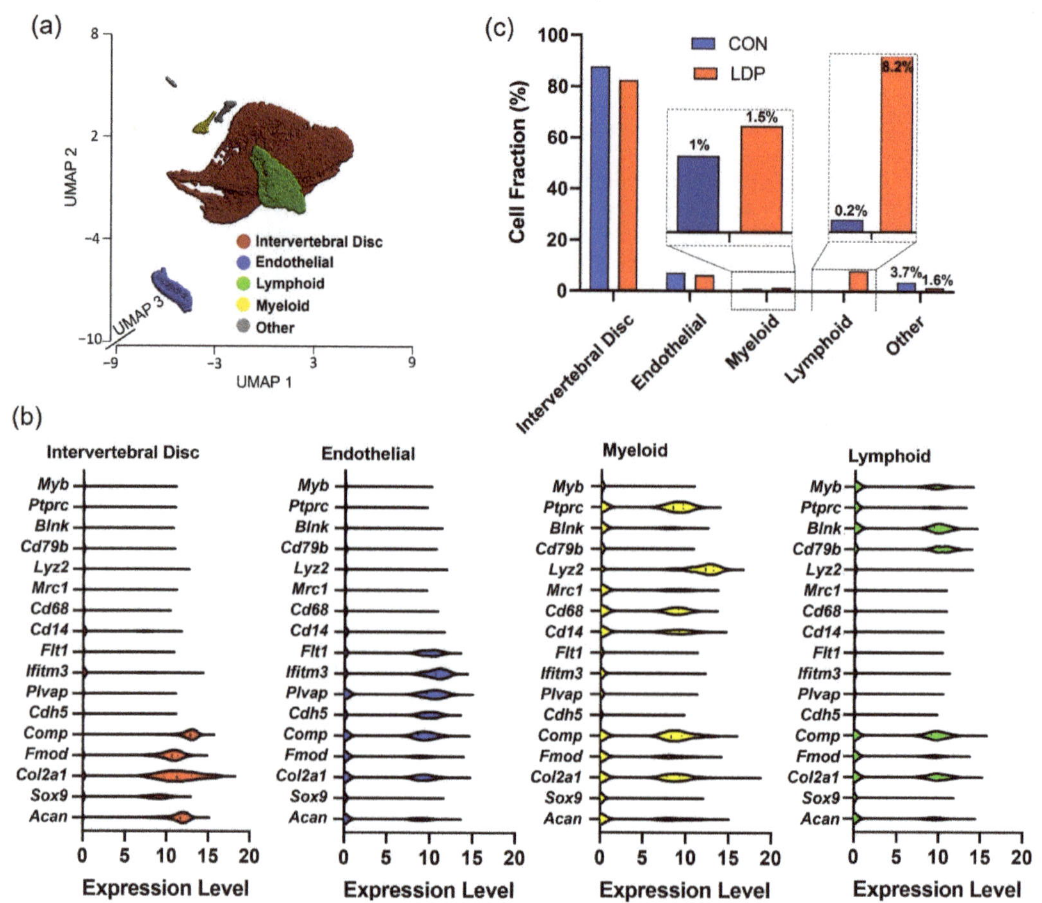

Figure 3. (**a**) UMAP plot representation of five cell types within the LDP group at two weeks after surgery. (**b**) The expression level of marker genes within intervertebral disc cells, endothelial cells, myeloid, and lymphoid cells in LDP groups at two weeks after surgery. (**c**) Percentage of all analyzed cells mapped to each cell type for CON and LDP groups at two weeks after surgery. Inset provided to highlight apparent differences in CON and LDP groups for immune cell populations.

Immune cell involvement in differences between LDP and CON gene expression patterns. We determined that immune cells were contributing to the upregulation genes due to LDP. Then, we classified the immune cell sub-types as either the myeloid or lymphoid clusters. There were significant increases in the number of cell mapping to the myeloid and lymphoid clusters for LDP samples as compared to CON samples at 2 weeks after surgery ($p < 0.00001$; Chi-square test), indicating an increase of infiltrating immune cells post-injury. The lymphoid cluster showed the largest difference in cell number between CON and LDP conditions, where the LDP samples had a cell count larger than 4.6× that of CON samples ($p < 0.00001$; Figure 5a). For the myeloid cells, we performed statistical analysis on the expression level of the pan-macrophage markers (*Cd4*, *Cd14*, *Cd68*, and *Lyz2*), M1 polarization (*Tlr2*, *Tlr4*, *Cd80*, and *Cd86*), and M2 polarization (*Arg1*, *Mrc1* (*Cd206*), *Mrs1* (*Cd204*)) cell surface markers (Figure 5b–d). The number of cells expressing pan-macrophage markers (*Cd4*, *Cd14*, *Cd68*, and *Lyz2*), as well as their respective levels of expression, indicated increased macrophages and heightened activity (Figure 5b). Furthermore, the number of cells expressing the canonical M1 cell surface markers *Tlr2*, *Tlr4*, and *Cd80* were not different between groups, but the number of *Cd86*+ cells ($p < 0.00001$) and their expression were higher in the LDP group (Figure 5c) [39]. Despite the similarity in the number of cells expressing M2 polarization markers between LDP and CON groups, including *Arg1*, *Mrc1* (*Cd206*), and *Mrs1* (*Cd204*), the expression levels were either indistinguishable (*Arg1*, *Cd206*) or lower (*Cd204*) compared with the CON group (Figure 5d) [39,40].

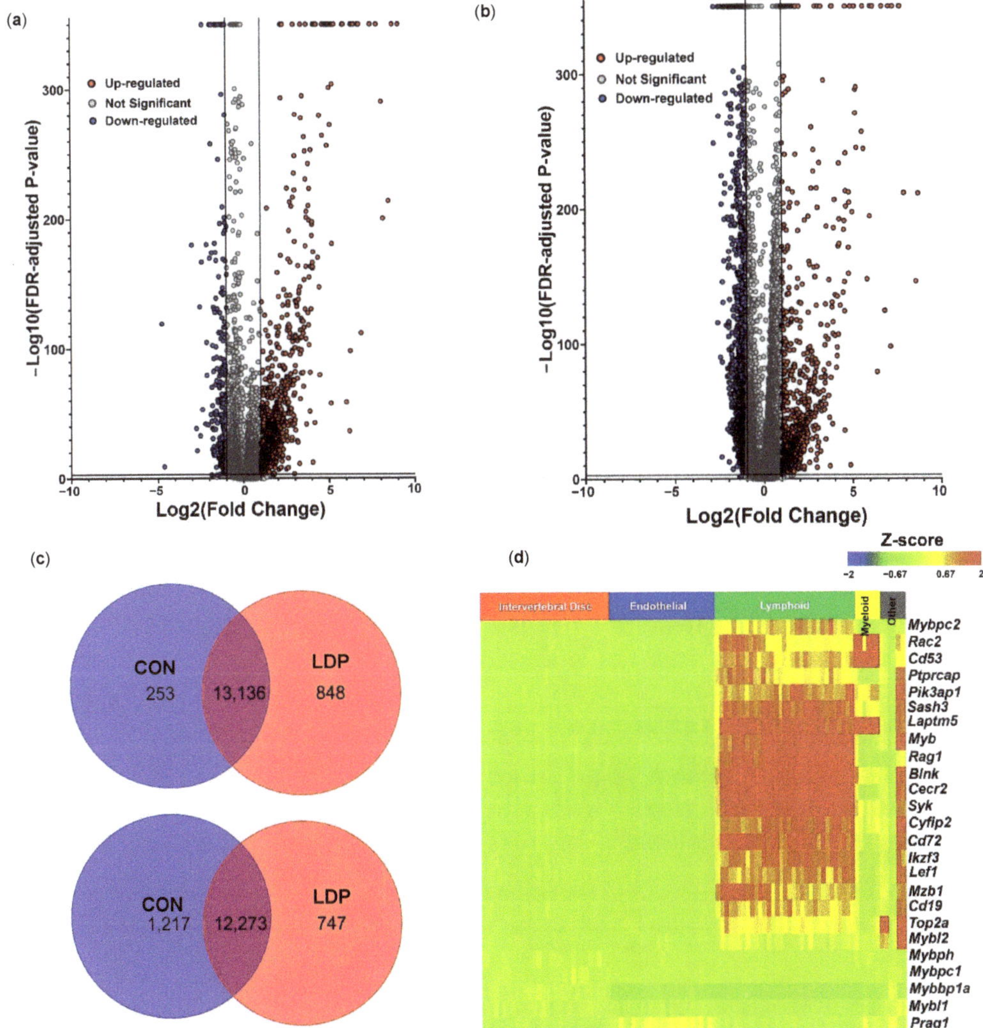

Figure 4. Volcano plot depicting specimen-specific differenti gene expression between CON and LDalP cells at two weeks after surgery. Two representative samples are shown—(**a**) sample 3 (left) and (**b**) sample 4 (right) (Red dots represent genes expressed at higher levels in the LDP group while blue dots represent genes with higher expression levels in CON group) (**c**) Venn diagram represents the total number of commonly upregulated and downregulated genes for LDP and CON cells in across multiple samples (i.e., samples 3 and 4 here). (**d**) Heatmap showing the pattern for a subset of commonly upregulated genes with LDP compared to CON groups as a function of cell grouping (*x*-axis).

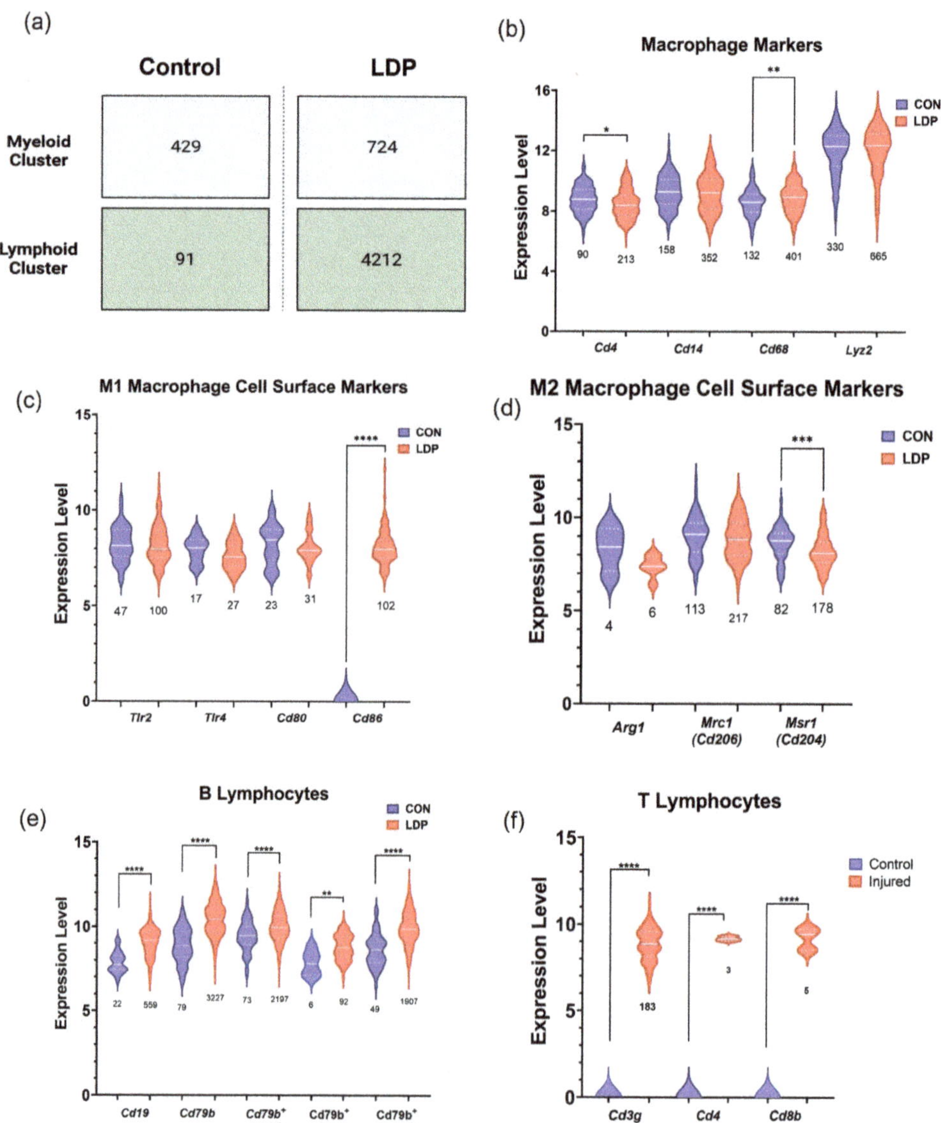

Figure 5. The identification of immune cell subtypes in the myeloid and lymphoid clusters. (**a**) Diagram showing the differences in cell numbers mapped to the myeloid and lymphoid clusters between the control (CON) and degenerated (LDP) samples at two weeks after surgery. (**b**) Pan-macrophage marker expression and cell count differences between CON and LDP groups were used to confirm the identification of cells in the myeloid cluster. Macrophage polarization was assessed by identifying the presence of either (**c**) M1 or (**d**) M2 polarization cell surface markers, and the number of cells that expressed them. The presence of (**e**) B and (**f**) T lymphocytes was identified in the lymphoid cluster by measuring pan and subtype-specific marker expression levels and cell count changes. Numbers underneath each plot denote the number of cells expressing each marker. A Student's t-test or a Mann–Whitney test was run to determine significance of marker expression levels. * = < 0.05, ** = <0.01, *** = <0.001, **** = <0.0001.

For the lymphoid clusters, we likewise performed statistical analysis for the B lymphocytes (*Cd19, Cd79b, Blnk, Cd38, Cxcr4*) and T lymphocytes (*Cd3g, Cd4, Cd8b*) markers by assessing the number of cells that expressed them (Figure 5e,f). B cell receptor and co-receptor, *Cd79B* and *Cd19* in LDP samples, showed greater levels of gene expression compared to CON cells, and were expressed in the majority of the lymphoid cells (Figure 5e). Markers of B lymphocyte subtypes, *Blnk, Cd38*, and *Cxcr4*, all showed significantly greater gene expression in LDP samples concurrent with large increases in cell numbers between CON and LDP at 2 weeks after surgery (Figure 5e). As for T lymphocytes, the gene expression and numbers of cells expressing pan T cell markers, *Cd3g*, the T cell receptor, were both significantly increased (Figure 5f). A small number of cells expressed *Cd4* or *Cd8b* subtype markers with significant increases in gene expression (Figure 5f).

Differential *Ngf* expression in IVD cells in LDP populations. The number and marker expression of cells mapping to the IVD subclusters were evaluated between CON and LDP populations at 2 weeks after surgery. The number of cells expressing *Ngf* and *Ngfr* (p75NTR) and their relative expression were used to compare the IVD subclusters including the NP, IAF and OAF (Figure 5a,b). The Gene ontology functions ranked from top to bottom by enrichment score derived from Fisher's exact test. The five most highly enriched processes identified by the Gene Set analyses were related to immune responses and associated regulatory processes (Figure 6c), consistent with the known infiltration of monocytes into the injured IVD post-surgical insult [41–43]. Genes associated with the top three regulated processes were upregulated in LDP samples and that are large drivers of the observed results are *Cd19, Cd37, Rag1, Blnk*, and *Myb*, as T- and B-cell associated marker genes.

While few differences were observed in relative expression levels of *Ngf* across the subclusters for the LDP group, the number of cells expressing *Ngf* increased following injury ($p < 0.00001$; Figure 6a). In CON cells from the NP cell sub-cluster, just 7.9% of cells expressed *Ngf* which increased to 12.9% following injury ($p < 0.00001$). *Ngf* is believed to be important for promoting neurite infiltration into the otherwise avascular and aneural tissues of the NP [38]. In OAF cells, the incidence of *Ngf* expressing cells increased from 1.0% of cells in CON samples at 2 weeks after surgery to 3.8% of cells in LDP samples, a greater than three-fold increase ($p < 0.00001$). Additionally, the gene expression of *Ngf* was significantly higher for all cells in the OAF subcluster of LDP samples ($p < 0.00001$). Likewise, the expression of *Ngfr*, which encodes for p75NTR—a low-affinity receptor for *Ngf*, was elevated in the OAF of LDP samples. Further, *Ngfr* was expressed in 2.3% of CON cells in the OAF sub-cluster increasing to 7.6% of cells expressed in the LDP injury group ($p < 0.00001$; Figure 6b).

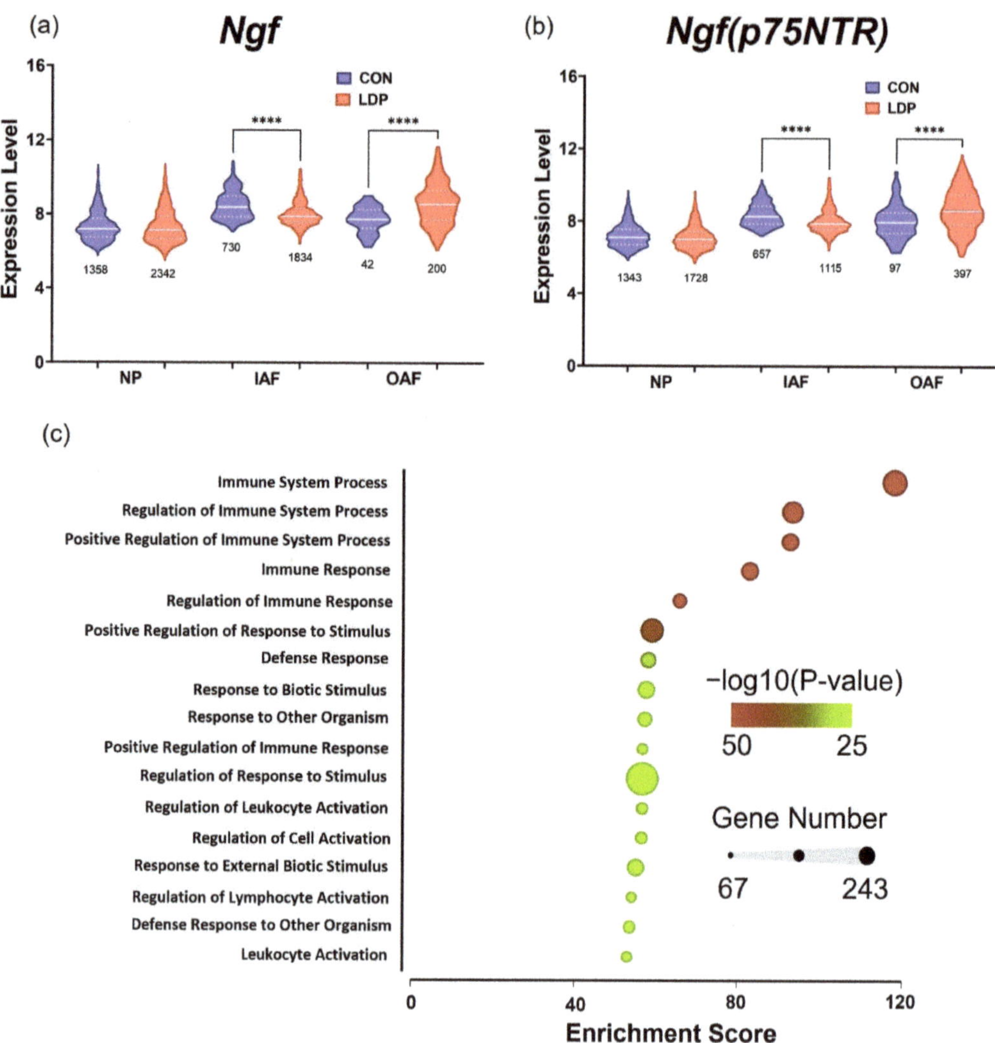

Figure 6. (**a**) *Ngf* and (**b**) *Ngfr* (p75NTR) expression and cell count counts for samples from CON and LDP populations at two weeks after surgery. Numbers underneath each plot denote the number of cells expressing each marker. A Student's *t*-test or a Mann–Whitney test was run to determine significance of marker expression levels. **** = <0.0001. (**c**) Representative plot of gene set enrichment analysis of upregulated genes corresponding to one LDP sample. Gene ontology functions are ranked from top to bottom by enrichment score derived from Fisher's exact test; The $-log_{10}^{p\ value}$ is represented by color where green denotes the highest level of statistical significance. The size of each node represents the number of genes in each gene ontology category.

3.2. Cell Populations Identified in IVD Tissues from 8 Weeks Post-Surgery

Samples in control IVD tissues. As described for samples harvested at 2 weeks post-surgery, unsupervised K-means clustering was applied to the CON group at 8 weeks post-surgery. The four identified clusters had a distribution of cell counts that was very similar across samples for the four CON samples at 8 weeks post-surgery (Figure 7a); note that the absence of a 5th identified cluster was related to a cluster (previously "endothelial cell cluster") with no mapped cells within. Violin plots of marker genes expression levels for all cells within each cluster were used to define the majority cell type in each cluster (Figure S1). The IVD cells contributed to most of the cluster (~83.7%) of all CON cells (8 weeks), followed by myeloid (~1.2%) and lymphoid cells (~8.4%). These differences from samples at 2 weeks post-surgery reflect a natural aging of the spine in the rat.

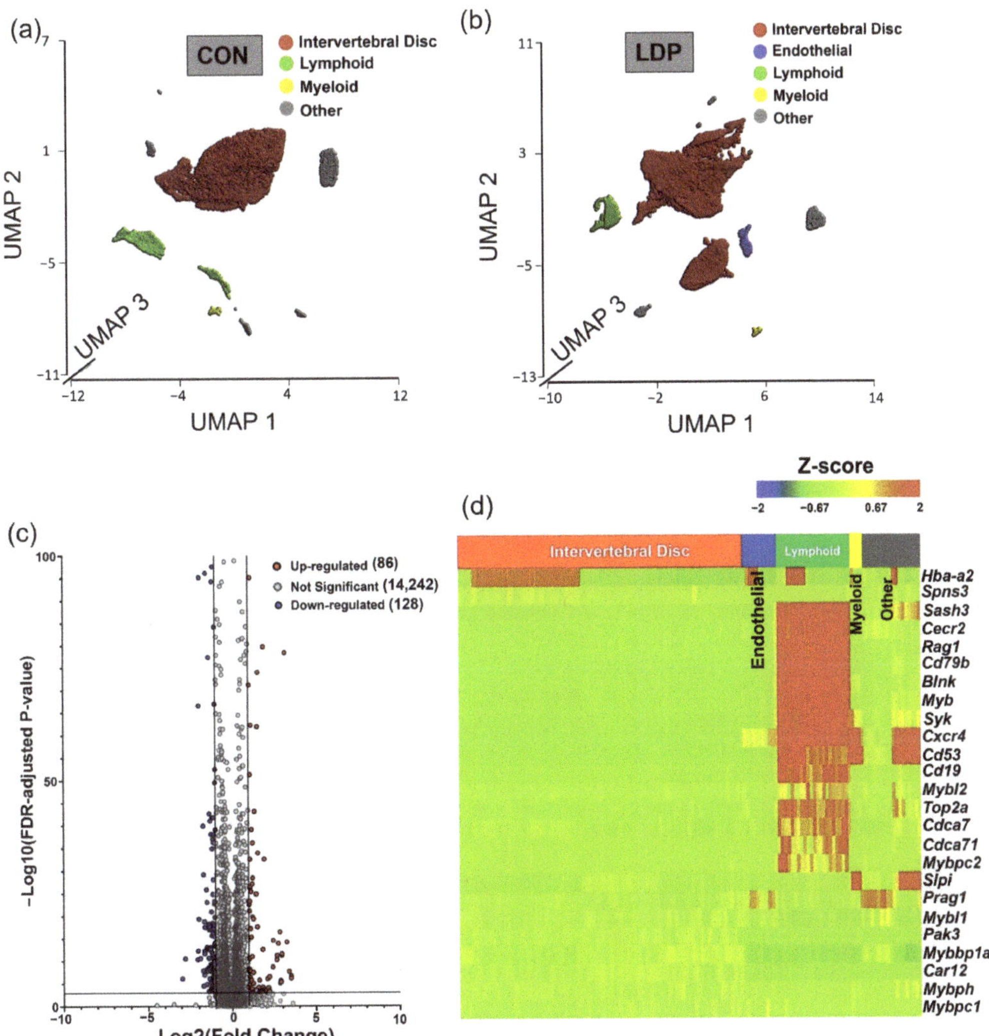

Figure 7. UMAP plot representation of all cell types within (**a**) CON and (**b**) LDP groups at 8 weeks after surgery. (**c**) Volcano plot depicting genes differentially expressed between sample-specific LDP relative to CON populations for one sample at 8 weeks after surgery (Red dots represent genes expressed at higher levels in LDP while blue dots represent genes with higher expression levels in CON group) (**d**) Heatmap showing the pattern for a subset of commonly upregulated genes with LDP compared to CON groups at 8 weeks after surgery as a function of cell grouping (x-axis).

Samples in degenerated IVD tissues. Samples harvested from the LDP segments of the rats at 8 weeks post-surgery underwent unsupervised K-means clustering as described above, with the observation that five clusters provided for optimal separation of the data (Figure 7b). The IVD cells contributed to most of the cluster (~87.4%), followed by endothelial cells (~2.3%), myeloid (0.8%) and lymphoid cells (~5.1%). Identification of cell populations within each cluster was based on known gene markers (Figure S2). These differences in sample numbers reflect the remodeling response to LDP at 8 weeks after surgery.

Differential gene expression between LDP and CON samples on a subject-specific basis. Differential gene expression between CON and LDP cells at 8 weeks after surgery demonstrated that 86 genes were significantly upregulated in LDP compared to CON, while 128 genes were more highly expressed in CON than in LDP samples (representative plot shown for sample #8 in Figure 6c). A subset of 25 genes was identified as commonly upregulated in LDP compared to CON samples and used to generate a heat map, showing that the majority of genes identified were associated with cells of the immune system, including *Blnk* and *Cd79b* used to identify immune cells, along with *Myb*, *Rag1*, and *Cecr2*. These findings point to similarities in upregulated genes at the 2 and 8 weeks timepoints after LDP surgery; however, the identification of a lower number of upregulated genes and lesser fold effect changes suggest that the dramatic differences observed at 2 weeks after surgery in LDP cells may reflect an acute response to the injury.

The heatmap shows that the upregulated genes mostly belong to the lymphocyte cluster (Figure 7d). Therefore, we explore the presence of B and T lymphocyte-specific markers within the lymphoid clusters. B cell receptor and co-receptor, *Cd79b* and *Cd19* in LDP samples, showed greater levels of gene expression compared to CON cells, and were expressed in the majority of the lymphoid cells (Figure 8b). *Blnk*, *Cd38*, and *Cxcr4* all showed significantly greater gene expression levels in LDP samples concurrent with large increases in cell numbers between CON and LDP at 8 weeks after surgery (Figure 8b). As for T lymphocytes, the only gene expression is related to *Cd3g* with zero expression level for *Cd4* and *Cd8b* in both CON and LDP (Figure 8c). These data suggest that the lymphocytes cluster is almost composed of B cells by less sign of T cells with respect to 2 weeks.

Gene set analyses revealed the five most highly significant processes identified point to B cell activation, receptor signaling, and differentiation (Figure 8a). Genes associated with the top three regulated processes that were found to be upregulated in LDP samples and that are large drivers of the observed results are *Cd19*, *Cd7bB*, *Rag1*, *Blnk*, *Dppr4*, *Il7r*, *Lef1*, and *Myb*.

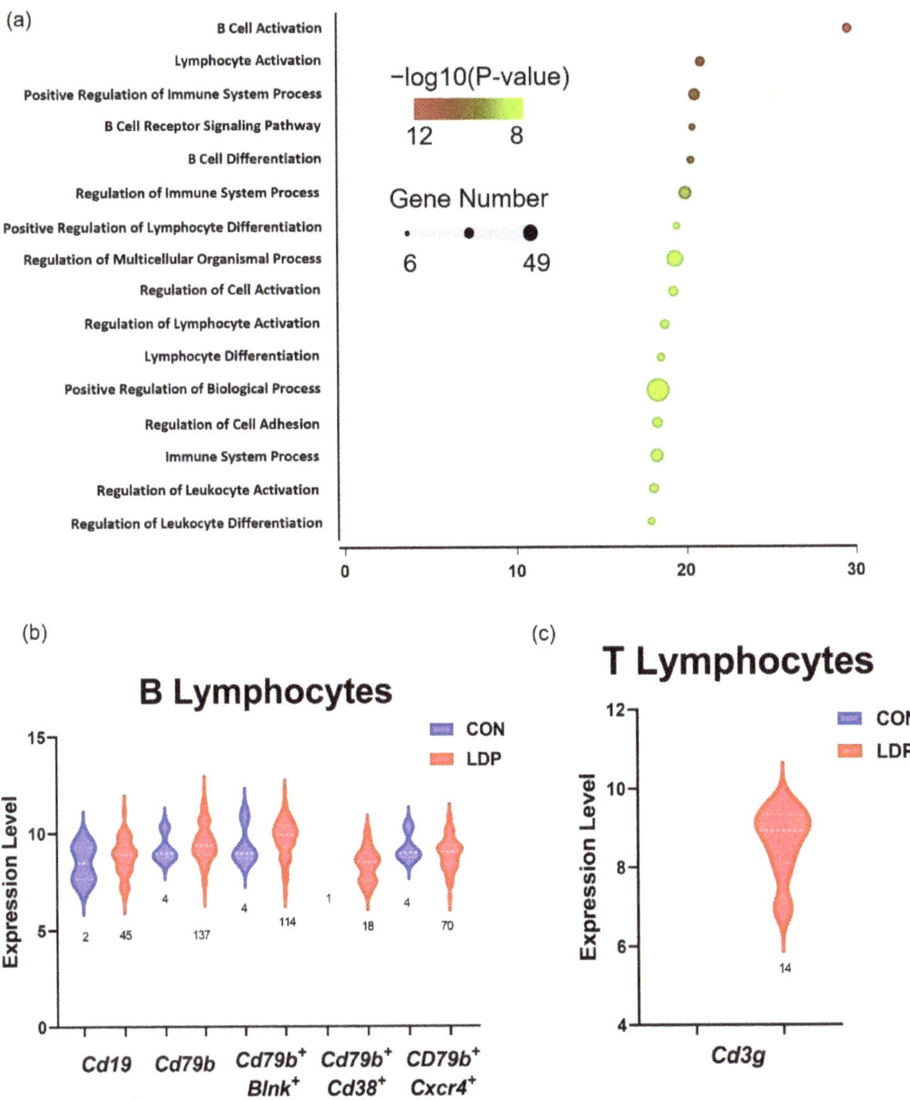

Figure 8. (**a**) Representative plot of gene set enrichment analysis of upregulated genes corresponding to one LDP sample. Gene ontology functions are ranked from top to bottom by enrichment score derived from Fisher's exact test; the $-log_{10}^{p\ value}$ is represented by color where green denotes the highest level of statistical significance. The size of each node represents the number of genes in each gene ontology category. (**b**) B and (**c**) T lymphocytes were identified in the lymphoid cluster of one rat sample taken at 8 weeks after surgery (sample #8) by measuring pan and subtype-specifi marker expression levels and cell count changes. Numbers below each plot denote the number of cells expressing each marker.

4. Discussion

Cells were isolated from the degenerated IVDs from a widely used rat model of lumbar disc puncture [24–28,44]. Single-cell RNA-seq was then performed on over 149,282 non-degenerate and degenerate IVD cells harvested from two timepoints following surgery to allow the progression of IVD degeneration. In brief, we identified four major cell types in the non-degenerate and degenerative IVD tissues ofc 2 weeks following surgery, based on the identification of genes previously identified as "classical" cell-specific markers [15]. For example, IVD cells are closely clustered around classical "chondrocyte-like" markers including *Acan*, *Col2a1*, *Sox9*, *Comp*, and *Fmod*. The cell classifications here in the master clustering and sub-clustering schemes for IVD cells from CON populations are consistent with a subset of those identified previously for NP cells; differences in cell identities were not noted between CON populations at 2 and 8 weeks following surgery, only differences in cell numbers mapping to each cluster.

Initial results that showed over 80% of all cells were identified as IVD cells at both 2 and 8 weeks, similar to a prior study showing that up to 99% of all sequenced cells mapped to a common transcriptomic profile [5]. The IVD cluster was then sub-clustered into NP, IAF, and OAF cells. For the markers chosen here for cell identification, the analysis revealed the absence of any singular genes that are specific to each subcluster. Rather, IVD cell type identification requires a set of genes that are either more or less highly expressed when compared to their spatial counterparts. Classical markers of IVD cells such as *Acan*, *Col1a1*, and *CoL2a1* and some NP-specific markers such as *Cd24* were useful to support the sub-clustering scheme; however, expression levels exhibited a continuous distribution from inner to outer IVD. Prior studies have sometimes attempted to anatomically isolate NP tissues from adjacent inner AF prior to performing sc-RNA-seq analyses, yet yield the same outcome [5,20,22,45], indicating the spatial overlapping nature of these tissues and the transition between cell populations in the IVD exists in a continuum. Instead, we found additional gene markers including *Igfbp5*, *Igfbp6*, *Lum*, *Myoc*, and *Fibin* useful as supplements of the "classical" markers used to separate outer AF from NP populations within the IVD [5].

In the current study, effort was taken to evaluate sample-specific variability in the clustering results for the CON populations at each time point (e.g., Figure 2c). Studies that pool multiple tissue levels across animals to procure sufficient cells for a single sample for RNA-sequencing may be susceptible to one sample disproportionately driving the clustering outcome. While we had differences in cell yields between rats, the clustering scheme was valued for its ability to produce repeatable results across animals (e.g., 82–92% of all cells mapping to IVD cluster. Approaches that define correlation analyses for clustering schemes as used by Calio and co-workers [16] would be a useful addition to all sc-RNA-seq analyses that employ multiple tissues from multiple subjects.

With LDP, differences were observed in the numbers of cells associated with each identified cluster. We observed that the LDP samples contained 1.5 times more myeloid cells than CON tissue and that macrophages were the most prominent immune subtype present in the myeloid cluster (Figure 3c; Figure 5b). Macrophages have been shown to infiltrate the herniated or injured disc in multiple species [23,42,46–50]. Accordingly, *Cd4* is a pan-macrophage marker in rats [51], *Cd68* is involved in antigen presentation [52,53], *Cd14* is essential for macrophage phagocytosis [54], and *Lyz2* is important for bacteriolysis [55]. In the LDP group, there were increased proportions of *Cd4+* ($p < 0.00001$), *Cd14+* ($p < 0.05$), *Cd68+* ($p < 0.00001$) and *Lyz2+* ($p = 0.05$) cells. Analysis of gene expression levels for each marker showed few differences in expression levels for these macrophage markers between CON and LDP groups, despite the higher numbers of cells expressing each marker in the LDP samples compared to CON (Figure 5b).

Macrophages can also undergo polarization into the classically activated pro-inflammatory phenotype, M1, or the alternative anti-inflammatory phenotype, M2 [39,56]. They have a remarkable range of cellular functions that can be elicited during tissue injury and repair dependent upon macrophage polarization within the affected tissue [57]. M1-type macrophages are classically activated and differentiation is stimulated by IFNγ to secrete pro-inflammatory cytokines such as IL6, IL12, and TNFα. M2-type macrophages are alternatively activated macrophages, and their differentiation is stimulation by IL4 produced by Th2 T lymphocytes. Their roles are to secrete anti-inflammatory cytokines and factors such as Arginase 1, IL10, and aid in wound healing, tissue repair, and regeneration. The myeloid cluster was highly enriched for both M1 macrophages and M2 macrophages at 2 weeks post IVD injury (Figure 5c,d). While both M1 and M2 macrophage markers are present, there is a significantly higher expression for *Cd86* (M1) and significantly lower expression for *Arg1* (M2) genes, suggesting a shift towards M1, or pro-inflammatory state with LDP at 2 weeks post-injury. This shift away from the M2 phenotype is consistent with the observations in the human degenerate nucleus pulposus cells that exhibited declining M2 proportion with increasing degeneration severity. Single-cell transcriptomic analyses of the human IVD cells show also the increasing activation of immune recruitment pathways during degeneration, demonstrating the robust interactions between the immune system and the degenerating IVD.

Our study here also identified the increased activation of B- and T-lymphocytes two weeks into the degenerative process. The majority of the top 25 commonly upregulated genes were in the myeloid and lymphoid clusters, and gene analysis led to the intriguing discovery that over 80% percent of those genes were highly enriched in lymphoid cell lineages. Notably, B cell-associate gene markers were the most represented group (40%: *Pika3p1, Rdac2, Blnk, Syk, Ikzf3, Mzb1, Cd19, Mybl1*) while the T cell-associated gene markers were slightly less prominent (25%: *Cyfip2*, Lef1, *Mybbp1a, Prag1, Myb*). Several gene markers were expressed in both T- and B- cells (35%: *Ptprcap, Sash3, Rag1, Top2a, Mybl2, Cd53, Laptm5*) (Figure 4d). B-lymphocyte specific markers included *Blnk, Rac2*, and *Pik3ap1*, which are involved in receptor signaling; *Syk, Cd72, Ikzf*, and *Cd19*, which are involved in differentiation and proliferation; and *Mzb1, Mybl1*, which are involved in antibody production and secretion. Due to the large presence of lymphocyte genes, we checked for the presence of B and T lymphocyte specific markers within the lymphoid clusters to confirm these immune cell types were present. B cell co receptor and receptor, *Cd19* and *Cd79b*, were present along with *Blnk*, which plays a role in B cell maturation; *Cd38*, which is expressed on B cell progenitors and plasma cells [58,59]; and *Cxcr4*, which plays a role in B cell activation and trafficking (Figure 5e) [60]. T lymphocytes cell receptor, *Cd3g*, had increased gene expression levels and an increase in cell number. Interestingly, only a small number of cells expressed *Cd4*, helper T cells, or *Cd8b*, cytotoxic T cells, emerged as identifiable in the LDP samples (Figure 5f).

Macrophages are capable of expressing cytokines that activate T-cells so that we examined T-cell specific markers for the incidence and genotype of T-cells. Notable T cell-specific markers were *Cyfip2*, needed for T cell adhesion; *Lef1*, required for IL17A expressing T cell maturation; and *Myb/Mybbp1a*, which are involved in T cell differentiation. Five genes were highly expressed in both B and T lymphocytes and regulate a range of functions such as cell adhesion and migration, *Cd53*; antigen receptor signaling, *Ptprcap* and *Sash3*; and VDJ recombination, *Rag1*, and *Top2a*. Though T-lymphocytes have been found in degenerating human IVDs [5], the presence of B-lymphocytes, which accounts for the majority of the highly expressed genes in the LDP population (e.g., *Cd72, Blnk, Syk*, and *Rag1*) has yet to be described during degeneration (Figure 5e,f). The involvement of immune cells and hematopoietic cells in the LDP tissue suggests that cell infiltration and differentiation were promoted by the surgical stab injury [14]. The role of lymphocytes after tissue injury had not been studied extensively; recent studies in bone fracture healing show clear roles for lymphocytes in mediating the tissue response after injury [61]. In bone fracture healing, lymphocytes were observed to infiltrate the injury site in a temporal fashion where they initially infiltrate the tissue early during the inflammation process, by 3 days post-injury, then leave and return after the peak inflammation window has passed and tissue proliferation has begun, around 14 days post-injury [61]. Lymphocytes function in the inflammation process by recruiting antigen-presenting cells, secreting cytokines to help mediate the inflammatory response, and secreting growth factors to aid in regeneration [61,62]. We also observed that lymphocytes persisted up to 8 weeks post injury (Figure 8a,b; Table 1). B-lymphocytes act as antigen-presenting cells and dampen the inflammatory response by promoting Treg survival and secreting IL10 [62]. The lymphoid cluster contained *Cd79b/Cd38* positive mature B-lymphocytes [59], and the presence of *Cd79b/Cxcr4* positive B-lymphocytes indicates activation and trafficking (Figure 5e,f) [60]. T-lymphocytes, in contrast, modulate antigen presenting macrophages, and regulate the inflammatory response with specific subtypes such as *Cd4* positive Tregs [63]. *Cd4* positive T-lymphocytes are associated with tissue repair after injury due to eliciting immunomodulatory effects, while cytotoxic *Cd8* positive T cells have been shown to have a negative effect on repair [61]. We observe *Cd3g* broadly expressed in T-lymphocytes, *Cd3g/Cd4* expression in T-lymphocytes helper cells, and *Cd3g/Cd8* expression in cytotoxic T-lymphocytes (Figure 5e,f). Our data here reveal a novel role for B-lymphocytes in concert with T-lymphocytes as part of the immune cascade during degeneration that persists in both our model here and in humans. Moreover, the immune markers, such as *Myb*, *Blnk*, and *Cd53*, remain similar over time for CON and LDP, suggesting that this is a sustained, progressive response due to degeneration rather than the acute trauma of the injury.

In addition to the described differences in cell number and gene expression between cells within the clusters of control and degenerative IVDs, we observed changes in *Ngf* levels and expression patterns for IVD cell sub-clusters with IVD degeneration. Both *Ngf* and *Ngfr* (encoding for p75NTR) were increased in expression levels in OAF cells, and the number of OAF cells expressing *Ngfr* was also increased. Additionally, NGF is pro-angiogenic that upregulates VEGF, and indirectly recruits endothelial cells [64]. Sustained NGF sensitizes innervating primary afferent neurons to produce hyperalgesia [65] via up-regulated NGF receptor expression in ganglion neurons [66], and promotes the expression of substance P, calcitonin gene-related peptide, and neuronal ion channels that may lead to long-term adaptive effects on nociceptors [67,68]. Moreover, NGF binding to its receptor can regulate collateral sprouting of sympathetic fibers to dorsal root ganglion neurons that may relate to the maintenance of sensitization or chronic pain. Indeed, a prior study of this LDP model in the rat revealed elevated protein expression for the NGF receptors TrkA and p75NTR in the innervating dorsal root ganglion neurons [44]. Secreted NGF can also drive the growth and differentiation of both B- and T-lymphocytes [69,70] in a pattern consistent with the elevated expression of these cell populations in the LDP samples.

A limitation of the current study is the absence of PCR confirmation of identified targets as performed in some prior studies [5,15,22]. As our interest was in numerically minor cell types involved in IVD degeneration, rather than phenotyping cells derived from discrete regions of the IVD, both immunolabeling and bulk-PCR proved difficult with remaining sample-specific tissues. Future work would be needed to design a surgical approach that made use of more lumbar levels, or other strategies such as flow cytometry, to support the collection of tissues and cells so that bulk-PCR or immunolabeling could be used to confirm the current findings. As is, the clear corroboration of clustering schemes that supported the use of "classical" cell-specific markers in the rat as used by Wang and co-workers [15], and the consistency observed in clustering results across four distinct rat subjects provides some measure of confidence and consistency to support the results provided here.

In summary, this study of single-cell RNA-seq in a rat model of IVD degeneration demonstrated changes in the numbers of cells associated with immune cell populations, with increased infiltration of B-lymphocyte and macrophage subsets into the IVD, particularly in early degeneration. While the surgical stab of the IVD undoubtedly provoked the release of soluble factors [71] that support endothelial cell infiltration, neurite invasion, and the activation of infiltrating monocytes that lead to neuronal sensitization and the identification of B- and T-lymphocytes as shown here, the results also point to a region-specific increase in the number of cells expressing *Ngf* and *Ngfr* (p75NTR) and associated transcriptional expression that may be one of the key findings that support the development of progressive IVD degeneration following stab injury to the IVD in this model and to annulus injury in the human. Most notably, there was a large presence of B-lymphocytes, which accounts for the majority of the highly expressed genes in the LDP population. These findings provide the basis for future work to investigate immune cell cross-talk and sub-type specificity during IVD degeneration.

Supplementary Materials: The following supporting information can be downloaded at: https://www.mdpi.com/article/10.3390/10.3390/app12168244/s1, Figure S1: Expression level of marker genes within intervertebral disc cells, myeloid, and lymphoid cells from CON samples at 8 weeks post-surgery. Figure S2: Expression level of marker genes within intervertebral disc cells, endothelial cells, myeloid, and lymphoid cells from LDP samples at 8 weeks post-surgery.

Author Contributions: M.R., J.E.S., S.W.C., G.W.D.E., D.S.P. and F.L. contributed to study design, and analyzed data. D.S.P., M.N.B., J.J.S. and L.J. performed the experiment and gathered the data. L.A.S., J.J.S. and S.Y.T. designed the research study and contributed to data interpretation. M.R., S.W.C., D.S.P., J.J.S., S.Y.T. and L.A.S. wrote the manuscript and contributed to data interpretation. All authors have read and agreed to the published version of the manuscript.

Funding: Funding was received from the National Institutes of Health (R01AR070975, R01AR078776, R01074441, T32EB028092), The Orthopedic Research and Education Foundation (resident research grant P20-03413), the National Science Foundation (Graduate Research Fellowship No. DGE-1745038), and the Rita Levi-Montalcini Postdoctoral Fellowship in Regenerative Medicine. We also acknowledge the Genome Technology Access Center (GTAC) at Washington University School of Medicine for help with genomic analysis (NIH P30 CA91842, NIH UL1TR002345). Figure 1 was created with Biorender.com (accessed on 1 June 2022).

Institutional Review Board Statement: The animal study protocol was approved by the Washington University Institutional Animal Care and Use Committee under Animal Welfare Assurance # A-3381-01 (Protocol # 20-0059).

Data Availability Statement: The RNA sequencing dataset generated during this study was deposited in the NCBI Gene Expression Omnibus repository (accession number: GSE211407).

Conflicts of Interest: The authors declare no conflict of interest.

References

1. Pattappa, G.; Li, Z.; Peroglio, M.; Wismer, N.; Alini, M.; Grad, S. Diversity of intervertebral disc cells: Phenotype and function. *J. Anat.* **2012**, *221*, 480–496. [CrossRef] [PubMed]
2. Virtanen, I.M.; Karppinen, J.; Taimela, S.; Ott, J.; Barral, S.; Kaikkonen, K.; Heikkilä, O.; Mutanen, P.; Noponen, N.; Männikkö, M.; et al. Occupational and Genetic Risk Factors Associated with Intervertebral Disc Disease. *Spine* **2007**, *31*, 1129–1134. [CrossRef] [PubMed]
3. Sakai, D. Future perspectives of cell-based therapy for intervertebral disc disease. *Eur. Spine J.* **2008**, *17*, 452–458. [CrossRef]
4. Noponen-Hietala, N.; Virtanen, I.; Karttunen, R.; Schwenke, S.; Jakkula, E.; Li, H.; Merikivi, R.; Barral, S.; Ott, J.; Karppinen, J.; et al. Genetic variations in IL6 associate with intervertebral disc disease characterized by sciatica. *Pain* **2005**, *114*, 186–194. [CrossRef] [PubMed]
5. Zhang, Y.; Han, S.; Kong, M.; Tu, Q.; Zhang, L.; Ma, X. Single-cell RNA-seq analysis identifies unique chondrocyte subsets and reveals involvement of ferroptosis in human intervertebral disc degeneration. *Osteoarthr. Cartil.* **2021**, *29*, 1324–1334. [CrossRef]
6. Sun, Y.; Lv, M.; Zhou, L.; Tam, V.; Lv, F.; Chan, D.; Wang, H.; Zheng, Z.; Cheung, K.; Leung, V. Enrichment of committed human nucleus pulposus cells expressing chondroitin sulfate proteoglycans under alginate encapsulation. *Osteoarthr. Cartil.* **2015**, *23*, 1194–1203. [CrossRef]
7. Sive, J.I.; Baird, P.; Jeziorsk, M.; Watkins, A.; Hoyland, J.A.; Freemont, A.J. Expression of chondrocyte markers by cells of normal and degenerate intervertebral discs. *Mol. Pathol.* **2002**, *55*, 91–97. [CrossRef]
8. Sakai, D.; Nakai, T.; Mochida, J.; Alini, M.; Grad, S. Differential Phenotype of Intervertebral Disc Cells: Microarray and Immunohistochemical Analysis of Canine Nucleus Pulposus and Anulus Fibrosus. *Spine* **2009**, *34*, 1448–1456. [CrossRef]
9. Kim, K.-W.; Lim, T.-H.; Kim, J.G.; Jeong, S.-T.; Masuda, K.; An, H.S. The Origin of Chondrocytes in the Nucleus Pulposus and Histologic Findings Associated with the Transition of a Notochordal Nucleus Pulposus to a Fibrocartilaginous Nucleus Pulposus in Intact Rabbit Intervertebral Discs. *Spine* **2003**, *28*, 982–990. [CrossRef]
10. Clouet, J.; Grimandi, G.; Pot-Vaucel, M.; Masson, M.; Fellah, H.B.; Guigand, L.; Cherel, Y.; Bord, E.; Rannou, F.; Weiss, P.; et al. Identification of phenotypic discriminating markers for intervertebral disc cells and articular chondrocytes. *Rheumatology* **2009**, *48*, 1447–1450. [CrossRef]

11. Chen, J.; Yan, W.; Setton, L.A. Molecular phenotypes of notochordal cells purified from immature nucleus pulposus. *Eur. Spine J.* **2006**, *15*, 303–311. [CrossRef] [PubMed]
12. Choi, K.-S.; Cohn, M.J.; Harfe, B.D. Identification of nucleus pulposus precursor cells and notochordal remnants in the mouse: Implications for disk degeneration and chordoma formation. *Dev. Dyn.* **2008**, *237*, 3953–3958. [CrossRef] [PubMed]
13. Tang, X.; Jing, L.; Chen, J. Changes in the Molecular Phenotype of Nucleus Pulposus Cells with Intervertebral Disc Aging. *PLoS ONE* **2012**, *7*, e52020. [CrossRef]
14. Tam, V.; Chen, P.; Yee, A.; Solis, N.; Klein, T.; Kudelko, M.; Sharma, R.; Chan, W.C.; Overall, C.M.; Haglund, L.; et al. DIPPER, a spatiotemporal proteomics atlas of human intervertebral discs for exploring ageing and degeneration dynamics. *eLife* **2020**, *9*, e64940. [CrossRef] [PubMed]
15. Wang, J.; Huang, Y.; Huang, L.; Shi, K.; Zhu, C.; Li, L.; Zhang, L.; Feng, G.; Song, Y. Novel biomarkers of intervertebral disc cells and evidence of stem cells in the intervertebral disc. *Osteoarthr. Cartil.* **2020**, *29*, 389–401. [CrossRef] [PubMed]
16. Calió, M.; Gantenbein, B.; Egli, M.; Poveda, L.; Ille, F. The Cellular Composition of Bovine Coccygeal Intervertebral Discs: A Comprehensive Single-Cell RNAseq Analysis. *Int. J. Mol. Sci.* **2021**, *22*, 4917. [CrossRef]
17. Fernandes, L.M.; Khan, N.M.; Trochez, C.M.; Duan, M.; Diaz-Hernandez, M.E.; Presciutti, S.M.; Gibson, G.; Drissi, H. Single-cell RNA-seq identifies unique transcriptional landscapes of human nucleus pulposus and annulus fibrosus cells. *Sci. Rep.* **2020**, *10*, 15263. [CrossRef]
18. McCann, M.R.; Patel, P.; Frimpong, A.; Xiao, Y.; Siqueira, W.L.; Séguin, C.A. Proteomic Signature of the Murine In-tervertebral Disc. *PLoS ONE* **2015**, *10*, e0117807.
19. Panebianco, C.J.; Dave, A.; Charytonowicz, D.; Sebra, R.; Iatridis, J.C. Single-cell RNA-sequencing atlas of bovine caudal intervertebral discs: Discovery of heterogeneous cell populations with distinct roles in homeostasis. *FASEB J.* **2021**, *35*, e21919. [CrossRef]
20. Gan, Y.; He, J.; Zhu, J.; Xu, Z.; Wang, Z.; Yan, J.; Hu, O.; Bai, Z.; Chen, L.; Xie, Y.; et al. Spatially defined single-cell transcriptional profiling characterizes diverse chondrocyte subtypes and nucleus pulposus pro-genitors in human intervertebral discs. *Bone Res.* **2021**, *9*, 37. [CrossRef]
21. Han, S.; Zhang, Y.; Zhang, X.; Zhang, H.; Meng, S.; Kong, M.; Liu, X.; Ma, X. Single-Cell RNA Sequencing of the Nucleus Pulposus Reveals Chondrocyte Differentiation and Regulation in Intervertebral Disc Degeneration. *Front. Cell Dev. Biol.* **2022**, *10*, 824771. [CrossRef] [PubMed]
22. Cherif, H.; Mannarino, M.; Pacis, A.S.; Ragoussis, J.; Rabau, O.; Ouellet, J.A.; Haglund, L. Single-Cell RNA-Seq Analysis of Cells from Degenerating and Non-Degenerating Intervertebral Discs from the Same Individual Reveals New Biomarkers for Intervertebral Disc Degeneration. *Int. J. Mol. Sci.* **2022**, *23*, 3993. [CrossRef] [PubMed]
23. Nerlich, A.G.; Weiler, C.; Zipperer, J.; Narozny, M.; Boos, N. Immunolocalization of Phagocytic Cells in Normal and Degenerated Intervertebral Discs. *Spine* **2002**, *27*, 2484–2490. [CrossRef] [PubMed]
24. Sobajima, S.; Shimer, A.L.; Chadderdon, R.C.; Kompel, J.F.; Kim, J.S.; Gilbertson, L.G.; Kang, J.D. Quantitative analysis of gene expression in a rabbit model of intervertebral disc degeneration by real-time polymerase chain reaction. *Spine J.* **2005**, *5*, 14–23. [CrossRef]
25. Kim, J.-S.; Kroin, J.S.; Li, X.; An, H.S.; Buvanendran, A.; Yan, D.; Tuman, K.J.; van Wijnen, A.J.; Chen, D.; Im, H.-J. The rat intervertebral disk degeneration pain model: Relationships between biological and structural alterations and pain. *Arthritis Res. Ther.* **2011**, *13*, R165. [CrossRef]
26. Lai, A.; Moon, A.; Purmessur, D.; Skovrlj, B.; Winkelstein, B.A.; Cho, S.K.; Hecht, A.C.; Iatridis, J.C. Assessment of functional and behavioral changes sensitive to painful disc degeneration. *J. Orthop. Res.* **2015**, *33*, 755–764. [CrossRef]
27. Miyagi, M.; Ishikawa, T.; Orita, S.; Eguchi, Y.; Kamoda, H.; Arai, G.; Suzuki, M.; Inoue, G.; Aoki, Y.; Toyone, T.; et al. Disk Injury in Rats Produces Persistent Increases in Pain-Related Neuropeptides in Dorsal Root Ganglia and Spinal Cord Glia but Only Transient Increases in Inflammatory Mediators: Pathomechanism of Chronic Diskogenic Low Back Pain. *Spine* **2011**, *6*, 26.
28. Elliott, D.M.; Yerramalli, C.S.; Beckstein, J.C.; Boxberger, J.I.; Johannessen, W.; Vresilovic, E.J. The Effect of Relative Needle Diameter in Puncture and Sham Injection Animal Models of Degeneration. *Spine* **2008**, *33*, 588–596. [CrossRef]
29. Risbud, M.V.; Schoepflin, Z.R.; Mwale, F.; Kandel, R.A.; Grad, S.; Iatridis, J.C.; Sakai, D.; Hoyland, J.A. Defining the phenotype of young healthy nucleus pulposus cells: Recommendations of the Spine Research Interest Group at the 2014 annual ORS meeting. *J. Orthop. Res.* **2014**, *33*, 283–293. [CrossRef]
30. Minogue, B.M.; Richardson, S.M.; Zeef, L.A.; Freemont, A.J.; Hoyland, J.A. Transcriptional profiling of bovine inter-vertebral disc cells: Implications for identification of normal and degenerate human intervertebral disc cell phenotypes. *Arthritis Res. Ther.* **2010**, *12*, R22. [CrossRef]
31. Rutges, J.; Creemers, L.; Dhert, W.; Milz, S.; Sakai, D.; Mochida, J.; Alini, M.; Grad, S. Variations in gene and protein expression in human nucleus pulposus in comparison with annulus fibrosus and cartilage cells: Potential associations with aging and degeneration. *Osteoarthr. Cartil.* **2010**, *18*, 416–423. [CrossRef] [PubMed]
32. Lee, C.R.; Sakai, D.; Nakai, T.; Toyama, K.; Mochida, J.; Alini, M.; Grad, S. A phenotypic comparison of intervertebral disc and articular cartilage cells in the rat. *Eur. Spine J.* **2007**, *16*, 2174–2185. [CrossRef] [PubMed]
33. Fujita, N.; Miyamoto, T.; Imai, J.-I.; Hosogane, N.; Suzuki, T.; Yagi, M.; Morita, K.; Ninomiya, K.; Miyamoto, K.; Takaishi, H.; et al. CD24 is expressed specifically in the nucleus pulposus of intervertebral discs. *Biochem. Biophys. Res. Commun.* **2005**, *338*, 1890–1896. [CrossRef]

34. Tang, X.; Jing, L.; Richardson, W.J.; Isaacs, R.E.; Fitch, R.D.; Brown, C.R.; Erickson, M.M.; Setton, L.A.; Chen, J. Identifying molecular phenotype of nucleus pulposus cells in human intervertebral disc with aging and degeneration. *J. Orthop. Res.* **2016**, *34*, 1316–1326. [CrossRef]
35. Yang, S.-H.; Hu, M.-H.; Wu, C.-C.; Chen, C.-W.; Sun, Y.-H.; Yang, K.-C. CD24 expression indicates healthier phenotype and less tendency of cellular senescence in human nucleus pulposus cells. *Artif. Cells Nanomed. Biotechnol.* **2019**, *47*, 3021–3028. [CrossRef] [PubMed]
36. Guo, W.; Zhang, B.; Li, Y.; Duan, H.Q.; Sun, C.; Xu, Y.Q.; Feng, S.Q. Gene expression profile identifies potential biomarkers for human intervertebral disc degeneration. *Mol. Med. Rep.* **2017**, *16*, 8665–8672. [CrossRef] [PubMed]
37. Brisby, H.; Olmarker, K.; Rosengren, L.; Cederlund, C.; Rydevik, B. Markers of Nerve Tissue Injury in the Cerebrospinal Fluid in Patients with Lumbar Disc Herniation and Sciatica. *Spine* **1999**, *24*, 742–746. [CrossRef]
38. Freemont, A.J.; Watkins, A.; Le Maitre, C.; Baird, P.; Jeziorska, M.; Knight, M.T.N.; Ross, E.R.S.; O'Brien, J.P.; Hoyland, J.A. Nerve growth factor expression and innervation of the painful intervertebral disc. *J. Pathol.* **2002**, *197*, 286–292. [CrossRef]
39. Porta, C.; Riboldi, E.; Ippolito, A.; Sica, A. Molecular and epigenetic basis of macrophage polarized activation. *Semin. Immunol.* **2015**, *27*, 237–248. [CrossRef]
40. Abdelaziz, M.H.; Abdelwahab, S.F.; Wan, J.; Cai, W.; Huixuan, W.; Jianjun, C.; Kumar, K.D.; Vasudevan, A.; Sadek, A.; Su, Z.; et al. Alternatively activated macrophages; a double-edged sword in allergic asthma. *J. Transl. Med.* **2020**, *18*, 1–12. [CrossRef]
41. Rousseau, M.-A.A.; Ulrich, J.A.; Bass, E.C.; Rodriguez, A.G.; Liu, J.J.; Lotz, J.C. Stab Incision for Inducing Intervertebral Disc Degeneration in the Rat. *Spine* **2007**, *32*, 17–24. [CrossRef] [PubMed]
42. Kanerva, A.; Kommonen, B.; Grönblad, M.; Tolonen, J.; Habtemariam, A.; Virri, J.; Karaharju, E. Inflammatory Cells in Experimental Intervertebral Disc Injury. *Spine* **1997**, *22*, 2711–2715. [CrossRef] [PubMed]
43. Risbud, M.V.; Shapiro, I.M. Role of cytokines in intervertebral disc degeneration: Pain and disc content. *Nat. Rev. Rheumatol.* **2013**, *10*, 44–56. [CrossRef]
44. Leimer, E.M.; Gayoso, M.G.; Jing, L.; Tang, S.; Gupta, M.C.; Setton, L.A. Behavioral Compensations and Neuronal Remodeling in a Rodent Model of Chronic Intervertebral Disc Degeneration. *Sci. Rep.* **2019**, *9*, 3759. [CrossRef] [PubMed]
45. Jiang, Y. Osteoarthritis year in review 2021: Biology. *Osteoarthr. Cartil.* **2021**, *30*, 207–215. [CrossRef]
46. Kawaguchi, S.; Yamashita, T.; Yokogushi, K.; Murakami, T.; Ohwada, O.; Sato, N. Immunophenotypic Analysis of the Inflammatory Infiltrates in Herniated Intervertebral Discs. *Spine* **2001**, *26*, 1209–1214. [CrossRef]
47. Doita, M.; Kanatani, T.; Harada, T.; Mizuno, K. Immunohistologic Study of the Ruptured Intervertebral Disc of the Lumbar Spine. *Spine* **1996**, *21*, 235–241. [CrossRef]
48. Doita, M.; Kanatani, T.; Ozaki, T.; Matsui, N.; Kurosaka, M.; Yoshiya, S. Influence of Macrophage Infiltration of Herniated Disc Tissue on the Production of Matrix Metalloproteinases Leading to Disc Resorption. *Spine* **2001**, *26*, 1522–1527. [CrossRef]
49. Grönblad, M.; Virri, J.; Tolonen, J.; Seitsalo, S.; Kääpä, E.; Kankare, J.; Myllynen, P.; Karaharju, E.O. A Controlled Immunohistochemical Study of Inflammatory Cells in Disc Herniation Tissue. *Spine* **1994**, *19*, 2744–2751. [CrossRef]
50. Shamji, M.F.; Setton, L.A.; Jarvis, W.; So, S.; Chen, J.; Jing, L.; Bullock, R.; Isaacs, R.E.; Brown, C.; Richardson, W.J. Proinflammatory cytokine expression profile in degenerated and herniated human intervertebral disc tissues. *Arthritis Rheum.* **2010**, *62*, 1974–1982.
51. Jefferies, W.A.; Green, J.R.; Williams, A.F. Authentic T helper CD4 (W3/25) antigen on rat peritoneal macrophages. *J. Exp. Med.* **1985**, *162*, 117–127. [CrossRef] [PubMed]
52. Chistiakov, D.A.; Killingsworth, M.C.; Myasoedova, V.A.; Orekhov, A.N.; Bobryshev, Y.V. CD68/macrosialin: Not just a histochemical marker. *Lab. Investig.* **2017**, *97*, 4–13. [CrossRef] [PubMed]
53. Song, L.; Lee, C.; Schindler, C. Deletion of the murine scavenger receptor CD68. *J. Lipid Res.* **2011**, *52*, 1542–1550. [CrossRef]
54. Tesfaigzi, Y.; Daheshia, M. *CD14. Encyclopedia of Respiratory Medicine*; Laurent, G.J., Shapiro, S.D., Eds.; Academic Press: Oxford, UK, 2006; pp. 343–347.
55. Markart, P.; Faust, N.; Graf, T.; Na, C.-L.; Weaver, T.E.; Akinbi, H.T. Comparison of the microbicidal and muramidase activities of mouse lysozyme M and P. *Biochem. J.* **2004**, *380*, 385–392. [CrossRef] [PubMed]
56. Lawrence, T.; Natoli, G. Transcriptional regulation of macrophage polarization: Enabling diversity with identity. *Nat. Rev. Immunol.* **2011**, *11*, 750–761. [CrossRef] [PubMed]
57. Mosser, D.M.; Edwards, J.P. Exploring the Full Spectrum of Macrophage Activation. *Nat. Rev. Immunol.* **2008**, *8*, 958–969. [CrossRef]
58. Shubinsky, G.; Schlesinger, M. The CD38 Lymphocyte Differentiation Marker: New Insight into Its Ectoenzymatic Activity and Its Role as a Signal Transducer. *Immunity* **1997**, *7*, 315–324. [CrossRef]
59. Oliver, A.M.; Martin, F.; Kearney, J.F. Mouse CD38 is down-regulated on germinal center B cells and mature plasma cells. *J. Immunol.* **1997**, *158*, 1108–1115.
60. Becker, M.; Hobeika, E.; Jumaa, H.; Reth, M.; Maity, P.C. CXCR4 signaling and function require the expression of the IgD-class B-cell antigen receptor. *Proc. Natl. Acad. Sci. USA* **2017**, *114*, 5231–5236. [CrossRef]
61. Könnecke, I.; Serra, A.; El Khassawna, T.; Schlundt, C.; Schell, H.; Hauser, A.; Ellinghaus, A.; Volk, H.D.; Radbruch, A.; Duda, G.N.; et al. T and B cells participate in bone repair by infiltrating the fracture callus in a two-wave fashion. *Bone* **2014**, *64*, 155–165. [CrossRef]
62. Muire, P.J.; Mangum, L.H.; Wenke, J.C. Time Course of Immune Response and Immunomodulation During Normal and Delayed Healing of Musculoskeletal Wounds. *Front. Immunol.* **2020**, *11*, 1056. [CrossRef] [PubMed]

63. D'Alessio, F.R.; Kurzhagen, J.T.; Rabb, H. Reparative T lymphocytes in organ injury. *J. Clin. Investig.* **2019**, *129*, 2608–2618. [CrossRef] [PubMed]
64. Skaper, S.D. Nerve growth factor: A neuroimmune crosstalk mediator for all seasons. *Immunology* **2017**, *151*, 1–15. [CrossRef] [PubMed]
65. Safieh-Garabedian, B.; Poole, S.; Allchorne, A.; Winter, J.; Woolf, C.J. Contribution of interleukin-1β to the inflamma-tion-induced increase in nerve growth factor levels and inflammatory hyperalgesia. *Br. J. Pharmacol.* **1995**, *115*, 1265–1275. [CrossRef]
66. Aoki, Y.; Takahashi, Y.; Ohtori, S.; Moriya, H.; Takahashi, K. Distribution and immunocytochemical characterization of dorsal root ganglion neurons innervating the lumbar intervertebral disc in rats: A review. *Life Sci.* **2004**, *74*, 2627–2642. [CrossRef]
67. Shu, X.; Mendell, L.M. Nerve growth factor acutely sensitizes the response of adult rat sensory neurons to capsaicin. *Neurosci. Lett.* **1999**, *274*, 159–162. [CrossRef]
68. Mukai, M.; Sakuma, Y.; Suzuki, M.; Orita, S.; Yamauchi, K.; Inoue, G.; Aoki, Y.; Ishikawa, T.; Miyagi, M.; Kamoda, H.; et al. Evaluation of behavior and expression of NaV1.7 in dorsal root ganglia after sciatic nerve compression and application of nucleus pulposus in rats. *Eur. Spine J.* **2013**, *23*, 463–468. [CrossRef]
69. Otten, U.; Ehrhard, P.; Peck, R. Nerve growth factor induces growth and differentiation of human B lymphocytes. *Proc. Natl. Acad. Sci. USA* **1989**, *86*, 10059–10063. [CrossRef]
70. Thorpe, L.W.; Perez-Polo, J.R. The influence of nerve growth factor on the in vitro proliferative response of rat spleen lymphocytes. *J. Neurosci. Res.* **1987**, *18*, 134–139. [CrossRef]
71. Lotz, J.C.; Ulrich, J.A. Innervation, Inflammation, and Hypermobility May Characterize Pathologic Disc Degeneration: Review of Animal Model Data. *J. Bone Jt. Surg.* **2006**, *88*, 76–82. [CrossRef]

applied sciences

Article

Influence of Angiopoietin Treatment with Hypoxia and Normoxia on Human Intervertebral Disc Progenitor Cell's Proliferation, Metabolic Activity, and Phenotype

Muriel C. Bischof [1], Sonja Häckel [2], Andrea Oberli [1], Andreas S. Croft [1], Katharina A. C. Oswald [2], Christoph E. Albers [2], Benjamin Gantenbein [1,2,*] and Julien Guerrero [1]

1 Tissue Engineering for Orthopaedics & Mechanobiology (TOM), Bone & Joint Program, Department for BioMedical Research (DBMR), Faculty of Medicine, University of Bern, CH-3008 Bern, Switzerland; muriel.bischof@gmx.ch (M.C.B.); andrea.oberli@dbmr.unibe.ch (A.O.); andreas.croft@dbmr.unibe.ch (A.S.C.); julien.guerrero@dbmr.unibe.ch (J.G.)

2 Department of Orthopaedic Surgery and Traumatology, Inselspital, Bern University Hospital, University of Bern, CH-3010 Bern, Switzerland; sonja.haeckel@insel.ch (S.H.); katharina.oswald@insel.ch (K.A.C.O.); christoph.albers@insel.ch (C.E.A.)

* Correspondence: benjamin.gantenbein@dbmr.unibe.ch

Citation: Bischof, M.C.; Häckel, S.; Oberli, A.; Croft, A.S.; Oswald, K.A.C.; Albers, C.E.; Gantenbein, B.; Guerrero, J. Influence of Angiopoietin Treatment with Hypoxia and Normoxia on Human Intervertebral Disc Progenitor Cell's Proliferation, Metabolic Activity, and Phenotype. *Appl. Sci.* **2021**, *11*, 7144. https://doi.org/10.3390/app11157144

Academic Editor: Ming Pei

Received: 12 July 2021
Accepted: 30 July 2021
Published: 2 August 2021

Publisher's Note: MDPI stays neutral with regard to jurisdictional claims in published maps and institutional affiliations.

Abstract: Increasing evidence implicates intervertebral disc (IVD) degeneration as a major contributor to low back pain. In addition to a series of pathogenic processes, degenerated IVDs become vascularized in contrast to healthy IVDs. In this context, angiopoietin (Ang) plays a crucial role and is involved in cytokine recruitment, and anabolic and catabolic reactions within the extracellular matrix (ECM). Over the last decade, a progenitor cell population has been described in the nucleus pulposus (NP) of the IVD to be positive for the Tie2 marker (also known as Ang-1 receptor). In this study, we investigated the influence of Ang-1 and Ang-2 on human NP cell (Tie2$^+$, Tie2$^-$ or mixed) populations isolated from trauma patients during 7 days in normoxia (21% O_2) or hypoxia ($\leq$5% O_2). At the end of the process, the proliferation and metabolic activity of the NP cells were analyzed. Additionally, the relative gene expression of NP-related markers was evaluated. NP cells showed a higher proliferation depending on the Ang treatment. Moreover, the study revealed higher NP cell metabolism when cultured in hypoxia. Additionally, the relative gene expression followed, with an increase linked to the oxygen level and Ang concentration. Our study comparing different NP cell populations may be the start of new approaches for the treatment of IVD degeneration.

Keywords: intervertebral disc (IVD); nucleus pulposus progenitor cells (NPPCs); angiopoietin-1 receptor (also known as Tie2); fluorescence-activated cell sorting; hypoxia; normoxia

1. Introduction

1.1. Low Back Pain and Intervertebral Disc Degeneration (IVDD)

Intervertebral disc degeneration is a significant and multifaceted cause of low back pain [1]. The so-called "discogenic pain" is described as a major cause of low back pain in 26–42% of cases [2]. Yet, it is known that the IVD undergoes degenerative changes earlier in life than do other tissues of major organ systems [1]. Whilst for the process of aging and degeneration, the IVD cells enter a state of cell senescence, showing an altered and often more catabolic metabolism [3], oftentimes the nucleus pulposus (NP) is primarily affected [3]. Moreover, the IVD possesses a very limited self-healing potential [4,5]. Conventional treatment strategies, such as analgetics, physical therapy, or surgical interventions, mainly address the symptoms of LBP, while no regeneration mechanisms in the IVD are activated [6]. The combination of tissue engineering strategies with stem cell-based therapies and approaches with native cells have gained significant momentum, being a desirable and promising novel treatment opportunity for degenerated IVDs since they aim to restore the properties of the IVD and, thus, their biological function [7].

1.2. Hypoxia vs. Normoxia in the Intervertebral Disc (IVD)

Another critical aspect in the IVD cell biology is that the NP is avascular, so its cell population does not receive a blood supply (nutrients diffuse through the vertebral endplates); therefore, oxygen levels in this central area are very low [8,9], causing a hypoxic environment [10]. As consequence, these NP cells may have adapted to these conditions. An increase in oxygen concentration during IVDD could be considered a pathological condition, unlike that observed in other tissues or wound healing [11].

1.3. Nucleus Pulposus Progenitor Cells (Also Known as Tie2+ Cells)

A cell source of current major interest is the tissue-specific Tie2 (also known as TEK receptor tyrosine kinase or angiopoietin-1 receptor or CD202b) positive NP progenitor cell (NPPC) population. These cells are perfectly adapted to the hypoxic environment and can therefore uphold their regenerative potency [12]. Notably, NPPCs are attributed with having potential in regenerative medicine because of their remarkable in vitro multipotency capability through their osteogenic, chondrogenic, and adipogenic lineages in addition to their self-renewal potential [13]. The harsh IVD microenvironment forms an obstacle to exogenous cells (such as bone marrow-derived mesenchymal stem cells) to survive and exhibit a tissue-regenerative function. Thus, NPPCs might pose a promising target to stimulate and induce tissue repair [14]. However, a major challenge in the NPPC therapy approach is caused by the decrease in the percentage of the Tie2+ NPPC population with aging and the extent of disc degeneration [1]. The methodology for optimized NP cell isolation (by cell sorting) and cell expansion are already described [6,15,16].

1.4. Tie2 Receptor and Its Ligands; Angiopoietin-1 and Angiopoietin-2

The receptor tyrosine kinase Tie2 is mainly expressed in endothelial cells and is essential for embryonal and adult angiogenesis and vascular maturation and maintenance [17,18] and its also expressed by early hematopoietic cells and subsets of monocytes express Tie2 [19]. Ang-1–4 are the growth factor ligands of the Tie2 receptor. In this project, we focused on Ang-1 and -2, which are mainly produced by vascular smooth muscle cells, throughout the entire vascular system, and endothelial cells respectively [20–22]. The binding of Ang-1 to its extracellular receptor Tie2 results in autophosphorylation and subsequent activation of downstream signaling pathways including the PI3-kinase/Akt [18,19,23]. The Tie2 receptor activation mediates stabilization, remodeling, and repair of the vascular system [18,24]. The inactivation of this pathway protects endothelial cells from apoptosis and inhibits inflammatory responses [18]. Cell-based Ang-1 gene therapy may represent a new strategy for treating various endothelium disorders, including several severe human pulmonary diseases [25,26]. Ang-2 fulfills a dual function by acting as an agonist or antagonist in a context-dependent manner [17]. while showing similar biological outcomes to Ang-1. However, it has to be emphasized that they are weaker in potency and kinetics. As a consequence, Ang-2, thus, may act as a partial Ang-1 agonist [17].

1.5. The Role of Tie2/Ang-1 and Ang-2 Signaling Pathway in Human NPCs and NPPCs

Sakai et al. [1] suggested that Tie2 is a sensitive marker of aging and the degeneration of IVDs, and therefore is strongly relevant in IVDD. In this study, the elementary role of the Tie2/Ang-1 signaling was already demonstrated. Endogenous Ang-1 provides an innate anti-apoptotic effect on human NP cells and is crucial for their survival. Additionally, enhancement of the proliferation of Tie2+ populations was achieved when NP cells were cultured with Ang-1. More recently, it was discovered that with aging and increasing degeneration, the imbalance between the Ang-2/Ang-1 ratio significantly rises [23]. Down-regulation by Ang can be associated with suppressing cell adhesion and viability and promoting the apoptosis of NPCs; therefore, the blockage of Ang-2 may represent another novel therapeutic approach to prevent NPC loss in IVD.

1.6. Hypothesis and Aims

This study aimed to investigate the effect of angiopoietin-1 and angiopoietin-2 in combination with culture under hypoxia and normoxia on three NP cell populations' (mixed, Tie2$^-$ and Tie2$^+$) proliferation, metabolism, and phenotype in the context of IVD regenerative medicine. The present work was undertaken to explore the hypothesis that, as NP tissue represents a heterogenous cell population, cells will react differently, in a dose-dependent effect to Ang-1 and Ang-2 stimulation. Secondly, we hypothesized that the variation in the measured parameters within those three cell populations is modified under normoxic and hypoxic conditions with or without stimulation with Ang-1 and Ang-2.

2. Materials and Methods

2.1. NPC Isolation

NP cells were isolated from IVD. The IVDs were obtained from human donors undergoing spinal surgery after trauma (Table 1). Written and informed consent was obtained from all participants. The tissue was dissected into nucleus pulposus tissue (NP), annulus fibrosus (AF) tissue, cartilaginous endplate (CEP) tissue, and cut into smaller fragments (0.3 cm^3). To avoid drying out the sample fragments, a sufficient amount of phosphate-buffered salt solution (PBS) was given over the tissue fragments.

Table 1. Overview of the sex and age of the patient when the sample was collected, IVD tissue location, cell's passage, and IVD status from all patients used in this project. Abbreviations: Th, thoracic; L, lumbar; S, sacrum; P, passage.

Donor Number	Sex	Age	IVD Level (Indication for Surgery)	Cell's Passage
1	Female	26	Th12/L1 (Trauma)	P3
2	Female	25	L1/L2 (Trauma)	P6
3	Male	20	L1/L2 (Trauma)	P3
4	Male	24	Th12/L1 (Trauma)	P4
5	Male	24	L3/L4; L4/L5; L5/S1 pooled (Trauma)	P6
6	Female	19	Th12/L1; L1/L2 pooled (Trauma)	P6
7	Female	20	Th12/L1; L1/L2 pooled (Trauma)	P2
8	Male	18	L1/L2 (Trauma)	P2

Subsequently, NP tissue samples were centrifuged for 5 min at 500× g (Eppendorf Centrifuge 5810, Vaudaux-Eppendorf, Schönenbuch, Switzerland). After centrifugation, PBS was removed and the samples were incubated with sterile 0.19% pronase solution (#10165921001; Roche Diagnostics, Basel, Switzerland) and digested on a shaker for 60 min at 12 rpm.

After one hour of digestion, tissue digested samples were centrifuged again for 5 min at 500× g. After removing the supernatant, the tissues were washed with PBS and digested with sterile collagenase type II (255 U/mg; #LS004174; Worthington Biochemical, Inc., London, UK). A solution of 0.025% collagenase type II was used, diluted in LG-DMEM (low glucose Dulbecco's Modified Eagle Medium (1 g/L; LG-DMEM; #31600-083; Thermo Fisher Scientific, Basel, Switzerland) +10% FBS (fetal bovine serum; #10500-064; Gibco Life Technologies, Basel, Switzerland). Then, the tissue samples were transferred into T75 flasks and left on a shaker at 12 rpm for overnight digestion.

The following day, cells and the digested tissues were filtered through a 100 µm strainer (#431752; Falcon; Becton, Dickinson and Company, Allschwil, Switzerland) to remove debris and washed once again with PBS. The cells were then resuspended in a LG-DMEM complete medium (CM) containing 0.22% sodium hydrogen carbonate (#31437-

500G-R; Sigma-Aldrich, Buchs, Switzerland), 10% FBS, 1 mM sodium pyruvate (#11360-039; Thermo Fisher Scientific), 1% penicillin/streptomycin/glutamine (#10378-016; P/S/G, 100 units/mL, 100 μg/mL and 292 μg/mL, respectively; Thermo Fisher Scientific), and 10 mM HEPES buffer solution (#15630-056; Thermo Fisher Scientific). The NP cell suspension was then transferred into T150 culture flasks (#90151; TPP, Trasadingen, Switzerland). Furthermore, cells from the AF and CEP tissues were frozen.

2.2. NPC Expansion

The cells were expanded with LG-DMEM (CM) supplemented with 2.5 ng/mL of FGF-2 (#100-18B, PeproTech, London, UK) [13]. The medium was changed twice per week. To obtain a sufficient number of cells for future experiments, NP cells were passaged until 100% confluency was reached, which corresponded to about 5–8 days of culture.

2.3. NPC Sorting

As soon as the NP cells reached 90% confluency, most of them were used to be sorted into a Tie2$^+$ cell population and a Tie2$^-$ cell population with fluorescence-activated cell sorting (FACS) as previously described [6,15,16]. In short, freshly isolated NPCs were incubated with an anti-human Tie2 PE-conjugated monoclonal mouse antibody (#FAB3131P, clone 83715, R&D systems) with 2 μL/10^6 cells for 30 min on ice and protected from light in 100 μL of FACS buffer ([PBS] with 0.1% bovine serum albumin (BSA; #A7030-100G; Sigma-Aldrich), P/S (100 units/mL and 100 μg/mL; Thermo Fisher Scientific), and 0.5 M ethylenediaminetetraacetic acid (EDTA; #03610; Fluka, Buchs, Switzerland)). Propidium iodide (Sigma-Aldrich) was used to exclude dead cells. Cell sorting was performed by FACS Diva III (BD Biosciences) based on the procedure previously reported [6]. The mouse IgG1 PE-conjugated antibody (#IC002P, clone 11711, R&D systems) was used as an isotype control to set the gate for sorting. The NP cell population prior to the sorting process was considered a mixed NP cell population, and the NP cell population after the sorting process was considered Tie2$^+$ or Tie2$^-$ NP cell populations.

2.4. NPC Seeding and Incubation with Ang-1 or Ang-2

In our study, 10^4 cells/cm^2 NPCs were seeded on 12-well plates (#92012, TPP) and incubated for 7 days, either within a control medium (CM; LG-MEM) or supplemented medium (LG-MEM with either angiopoietin-1 or angiopoietin-2 (#130-06 and #130-07, respectively, Peprotech) at a concentration of 10 ng/mL, 50 ng/mL, and 100 ng/mL). For each donor, the plates were prepared as duplicates, such that one plate per condition could be tested under a normoxic condition (21% of O_2) and the second plate was tested under 2% O_2 hypoxia (Invivo2 400 #0612NCF05, Ruskinn Technology Ltd., London, UK).

2.5. Resazurin Sodium Salt Cell Activity Assay

The resazurin sodium salt (#R7017-1G, Sigma-Aldrich) assay was performed at a concentration of 10 μg/mL added to the respective medium for 4 h at 37 °C under normoxic or hypoxic conditions, respectively. All test conditions were run in triplicate (#655101_100, Greiner Bio-One, St. Gallen, Switzerland). The fluorescence signal was quantified with a SpectraMax M5 (Molecular Devices, distributed by Bucher Biotec, Switzerland) microplate reader with excitation and emission wavelengths of 560 nm and 600 nm, respectively. The resulting values were normalized to the respective day 7 control sample CM without any added growth factors and incubated in normoxia or hypoxia, respectively.

2.6. Papain Digestion and DNA Quantification

The quantity of DNA present in each condition was measured for each sample after overnight papain digestion. The absolute amount of DNA was determined, using the CyQUANT^{TM} Cell Proliferation Assay Kit (#7026 Molecular Probes, Fisher Scientific) according to the manufacturer's instruction. All test groups were analyzed in triplicate.

2.7. RNA Extraction and cDNA Synthesis

The total RNA was isolated from monolayer cultures using the GenEluteTM Mammalian Total RNA Miniprep Kit (#RTN70-1KT; Sigma-Aldrich), according to the manufacturer's instruction. The following four steps were performed: releasing of the RNA from cells, binding of the RNA to the affinity column, washing to remove contaminants, and elution of the purified RNA. The purity of the RNA was assessed by measuring the A260/A280 ratio.

Reverse transcription of total RNA into double-stranded cDNA was performed with the High Capacity Reverse Transcription cDNA Kit (Thermo Fisher Scientific, Inc., cat #4368814), including 10X RT Buffer, 10X Random Primers, 25X dNTP Mix, and MultiScribe Reverse Transcriptase (50 U/mL), which were combined into a mastermix with additional RNase-free water and the addition of about 500 ng of total RNA.

2.8. Quantitative Polymerase Chain Reaction (qPCR)

To study the relative gene expression, cDNA (2 µL) was mixed with the PCR reaction solution (2X SYBR green master mix, #B21202; Bimake) containing 0.25 µM specific primers (Table 2). A real-time quantitative polymerase chain reaction (qPCR) was performed using the CFX96TM Real-Time System (#185-5096 C1000 Touch Thermal Cycler; Bio-Rad Laboratories) under standard thermal conditions. The qPCR program consisted of a two-step cycling protocol: an initial denaturation at 95 °C for 5 min, followed by 39 PCR cycles at 95 °C for 10 sec, and an annealing temperature of 61 °C for 30 sec. To control the specificity of the amplification products, a melting curve analysis was carried out for each reaction. The results were expressed relative to gene expression levels on day 0. Gene expression was calculated by the $2^{-\Delta\Delta Ct}$ method [27], and the results were normalized to the reference genes glyceraldehyde-3-phosphate dehydrogenase (GAPDH) and 18S ribosomal RNA (18S) expression. All reactions were performed in triplicate.

Table 2. Overview of human genes used for qPCR using glyceraldehyde-3-phosphate dehydrogenase (*GAPDH*) and *18S* ribosomal RNA (*18S*) as a reference gene. Genes analyzed for the NPCs: receptor tyrosine kinase (*TEK*), collagen type II (*COL2*), hypoxia-inducible factor-1α (*HIF1A*), aggrecan (*ACAN*). (F = forward, R = reverse).

Gene	Full Gene Name	Primer Nucleotide Sequence from 5′ to 3′
GAPDH	Glyceraldehyde-3-Phosphate Dehydrogenase	F—ATC TTC CAG GAG CGA GAT R—GGA GGC ATT GCT GAT GAT
18S	18S ribosomal RNA	F—CGA TGC GGC GGC GTT ATT R—TCT GTC AAT CCT GTC GTC CGT GTC C
TEK	TEK Receptor Tyrosine Kinase	F—TTA GCC AGC TTA GTT CTC TGT GG R—AGC ATC AGA TAC AAG AGG TAG GG
COL2	Collagen Type II	F—AGC AGC AAG AGC AAG GAG AA R—GTA GGA AGG TCA TCT GGA
HIF1A	Hypoxia Inducible Factor 1 Subunit Alpha	F—GTC GCT TCG GCC AGT GTG R—GGA AAG GCA AGT CCA GAG GTG
ACAN	Aggrecan	F—CAT CAC TGC AGC TGT CAC R—AGC AGC ACT ACC TCC TTC

We first assessed the effect of Ang-1/2 at different concentrations in hypoxia and normoxia on the relative expression of two genes that are associated with ECM production in the IVD, namely aggrecan (ACAN). We also examined the HIF1A gene, which is the crucial mediator of the adaptive response of cells to hypoxia. Furthermore, to estimate the expression of the Tie2 receptor, regarding the effect of the Ang-1/2 proteins and the microenvironments (normoxia vs. hypoxia) in the NP cells mixed population, and the NP Tie2^{-} and NP Tie2^{+} cell populations, we analyzed the TEK gene.

2.9. Statistical Analysis

Statistics were performed using GraphPad Prism (version 8.0.1 for Windows, GraphPad Software; San Diego, CA, USA). All data were initially tested for normal distribution using the Shapiro–Wilk test. Due to lack of normality a nonparametric distribution was then assumed. Statistical significance was determined using the Kruskal–Wallis's test followed by the Sidak's multiple comparison test. Results were presented as mean ± standard deviation (SD). A p-value < 0.05 was considered significant. Asterisks within figures denote the degree of statistical relevance observed: * < 0.05; ** < 0.01; *** < 0.001; **** < 0.0001.

3. Results

3.1. Effect of Oxygen Tension and Angiopoietin-1/2 on NP Cell's Proliferation

To explore the role of Ang-1 and Ang-2 in cell proliferation on the three NP cells populations (namely, Tie2$^+$, Tie2$^-$and mixed populations) we assessed the amount of DNA under normoxic and hypoxic conditions for the NP cells mixed, NP Tie2$^-$ cells, and NP Tie2$^+$ cells populations (Figure 1).

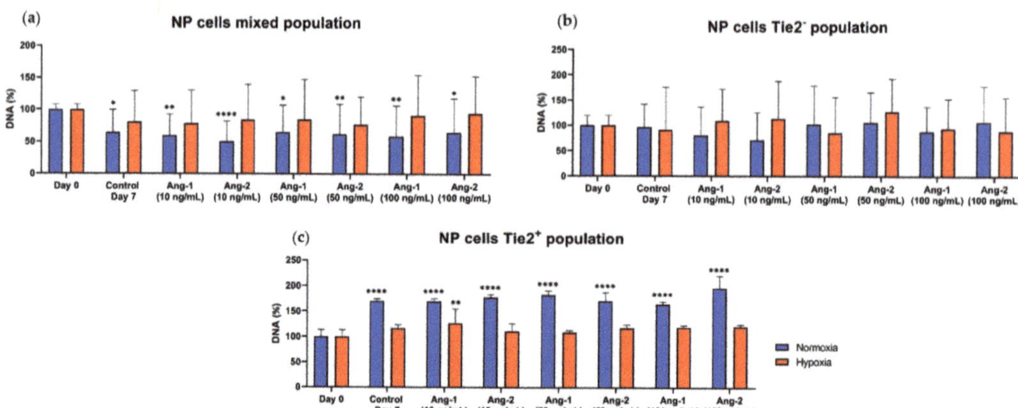

Figure 1. Amount of DNA measured of NP cells populations under normoxia and hypoxia assessed by Hoechst DNA assay on Day 7. Ang-1 and Ang-2 were administrated in three concentrations: 10, 50 and 100 ng/mL. The control Day 0 samples were used as reference. (**a**) Amount of DNA measured of NP cells mixed population (N = 7); (**b**) amount of DNA measured of NP cells Tie2$^-$ population (N = 3); (**c**) amount of DNA measured of NP cells Tie2$^+$ population (N = 2). Bars represent mean ± SD. A p-value < 0.05 was considered significant. Asterisks within figures denote the degree of statistical relevance observed: * < 0.05; ** < 0.01; **** < 0.0001.

Concerning the NP cells mixed population (Figure 1a), under normoxic conditions, the control Day 7 and the stimulated samples showed a significant decrease in their amount of DNA in a range of 35 to 47%, compared to the control Day 0 sample. However, all the samples did not show any significant differences in their amount of DNA, compared to the Day 0 control sample in hypoxia. Moreover, over the whole NP cells mixed population, no concentration-dependent or protein-dependent trends were apparent.

In the NPCs Tie2$^-$ population (Figure 1b), a slight upregulation in the amount of DNA was observed in every sample, in normoxia and hypoxia, compared to the Day 0 sample. In normoxia, the amount of DNA was increased up to 1.5 folds and in hypoxia, it was increased 1.25- to 1.6-fold. However, no concentration-dependent or O_2-dependent effect or trend was detectable for any sample condition, treated or either untreated with Ang-1 or Ang-2.

The entire collection of samples treated in the NP cells Tie2$^+$ population (Figure 1c) showed an increased amount of DNA, compared to the Day 0 sample. However, no concentration-dependent stimulation or protein-dependent effect was observable either in normoxia. In the hypoxic condition, only the sample treated with Ang-1 at (10 ng/mL) showed a significant increase in their amount of DNA compared to the control.

3.2. Effect of Oxygen Tension and Angiopoietin-1/2 on NP Single Cell's Metabolism

Further, we examined the metabolic activity/DNA ratio at Day 7, which is the most relevant parameter, as it refers to single-cell metabolic activity (Figure 2).

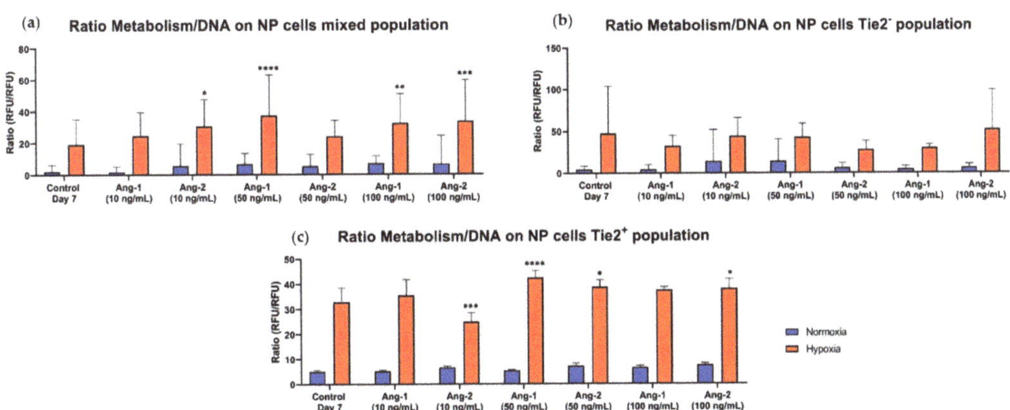

Figure 2. Metabolic activity on single-cell level of NP cells populations in normoxia and hypoxia on Day 7. Ang-1 and Ang-2 were administrated in three concentrations: 10, 50 and 100 ng/mL. The control Day 7 samples were used as reference. (**a**) Metabolic activity on single-cell level of NP cells mixed population; (**b**) metabolic activity on single-cell level of NP cells Tie2$^-$ population; (**c**) metabolic activity on single-cell level of NP cells Tie2$^+$ population. Bars represent mean ± SD. A p-value < 0.05 was considered significant. Asterisks within figures denote the degree of statistical relevance observed: * < 0.05; ** < 0.01; *** < 0.001; **** < 0.0001.

In the NP mixed population under normoxic conditions, the stimulated samples showed a 2- to 3-fold increase in the metabolism/DNA ratio, compared with the Day 7 control sample. In summary, for normoxia, a stimulation effect was detectable, but not specifically in concentration- or protein-dependency (Figure 2a). Furthermore, in hypoxic conditions, for the NP mixed population, the stimulated samples all presented an increase in their ratio metabolism/DNA, as they were 4 to 8 times higher than in normoxic conditions. Hence, no clear trend in concentration- or protein-dependency was observable. Concerning the Ang-1 stimulated samples, the metabolism/DNA ratio increased from 10 to 50 ng/mL concentrations and then decreased for the 100 ng/mL sample, with a significant increase for Ang-1 at 50 and 100 ng/mL. Regarding the Ang-2 treated samples, the metabolism/DNA ratio decreased from 10 to 50 ng/mL and then increased again for the Ang-2 for the 100 ng/mL treated sample, with a significant increase for Ang-2 at 10 and 100 ng/mL.

For the NP Tie2$^-$ cells population under normoxic condition, the ratios of the Ang-2 at 10 ng/mL and the Ang-1 at 50 ng/mL samples were 3- to 3.5-fold increased, compared to the Day 7 control sample (Figure 3b). For the other Ang-1 or Ang-2 treated samples, the ratios stayed at an equal level to the Day 7 control sample, or they even decreased. In hypoxia, all the stimulated samples showed a decrease in their metabolism/DNA ratio, except for the Ang-2 at the 100 ng/mL sample, whose ratio stayed at a similar level to the Day 7 control sample. as Additionally, for the mixed population, no trend in the effect of the concentration, nor the protein, was assessable. The Ang-1 treated samples presented an increase in their metabolism from the 10 to 50 ng/mL samples and then a decrease to the 100 ng/mL sample. The Ang-2 treated samples showed a decrease from the 10 ng/mL sample to the 50 ng/mL sample and then an increase to the 100 ng/mL sample. However, no significant increase or decrease was detected.

In the NP Tie2$^+$ cells population, we observed a similar ratio for all the samples in normoxia, either stimulated with Ang-1/2 or not (Figure 2c). The same picture was present in hypoxia. All samples showed similar or slightly elevated ratios, compared to the day 7 control sample, except for the Ang-2 at the 10 ng/mL sample, which significantly decreased in its ratio, compared to the control sample. Therefore, a significant increase in stimulation with Ang-1 at 50 ng/mL and Ang-2 at 10, 50 and 100 ng/mL was detected.

3.3. Effect of Oxygen Tension and Angiopoietin-1/2 on ECM Related-Genes

3.3.1. Analysis of Aggrecan Relative Gene Expression

Concerning the relative gene expression of *ACAN* for the NP cells mixed population, we did not observe any significant difference for normoxia, compared to the control condition (Figure 3a). No effects were detected in a dose-dependent manner for angiopoietin on the relative gene expression nor for specific variation in the use of Ang-1 or Ang-2. Nonetheless, all treated samples, either with Ang-1 or Ang-2, showed an equal amount of *ACAN* as the control samples, or at least a 2- to 3-fold increase in their ACAN relative expression. However, Ang-2 samples at 100 ng/mL in normoxia presented a downregulation in their *ACAN* gene expression.

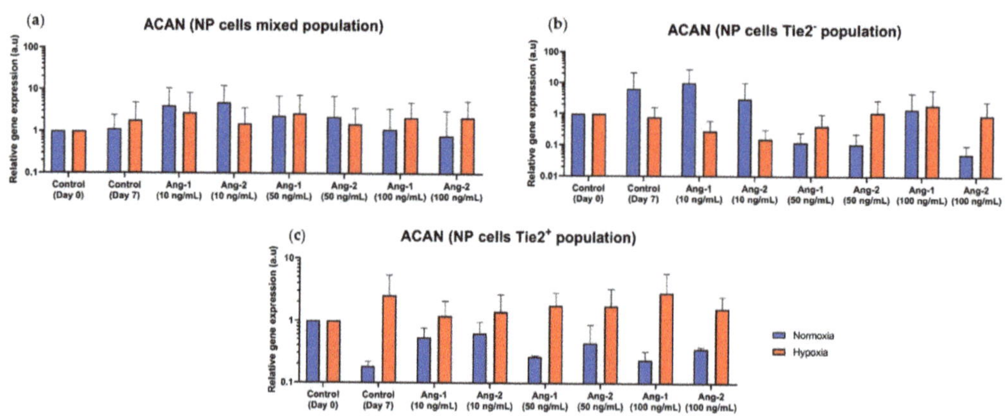

Figure 3. Relative gene expression of the three NP cells populations, comparing the effect of Ang1/2 at different concentrations in normoxia and hypoxia on the ACAN gene expression. The control (Day 0) samples were used as reference (set at 1). Ang-1 and Ang-2 were administrated in three concentrations: 10, 50 and 100 ng/mL. (**a**) ACAN relative gene expression of NP cells mixed population; (**b**) ACAN relative gene expression of NP cells Tie2$^-$ population; (**c**) ACAN relative gene expression of NP cells Tie2$^+$ population. Bars represent mean $\pm$ SD.

Like the NP cells mixed population, the control Day 0 and control Day 7 samples showed non-significant differences in their relative expression of *ACAN* in normoxia and hypoxia for the NP cells Tie2$^-$ population (Figure 3b). However, a remarkable effect in a dose-dependent manner was detectable for the Ang-1/2 sample at 10 ng/mL in normoxia, the Ang-1 sample at 100 ng/mL in normoxia and hypoxia, and the Ang-2 sample at 100 ng/mL in hypoxia.

The NP cells mixed, NP Tie2$^-$ cells, and NP Tie2$^+$ cells populations showed no differences in the gene expression of *ACAN*, compared to the control Day 0 (Figure 3c). We detected no specific variation in the use of Ang-1 or Ang-2 in normoxia and hypoxia but detected a dose-dependent increase in relative gene expression for hypoxia. In normoxia, we detected no effect in a dose-dependent manner of angiopoietin on the relative gene expression or specific variation in the use of Ang-1 or Ang-2. All the stimulated samples in normoxia were downregulated in their *ACAN* gene expression by 40 to 75%, compared to the control Day 0 sample.

3.3.2. Analysis of Collagen Type II Relative Gene Expression

The relative gene expression of *COL2* for every sample in normoxia and hypoxia increased compared to the control Day 0 sample (Figure 4a). The peak values in *COL2* relative gene expression compared to the control Day 0 sample were reached on a significant level in the Ang-1 and Ang-2 at 10 ng/mL treated samples in normoxic condition. All stimulated hypoxia samples and the control Day 7 sample showed a 2- to 2.5-fold increase in their *COL2* relative gene expression. Thus, we detected no effect in a dose-dependent manner of angiopoietin on the relative gene expression or specific variation in the use of Ang-1 or Ang-2 in hypoxia.

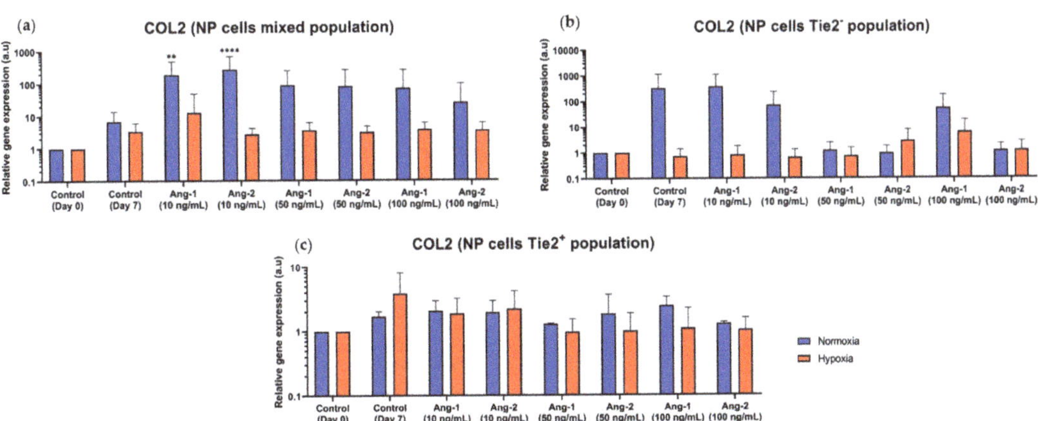

Figure 4. Relative gene expression of the three NP cells populations, comparing the effect of Ang1/2 at different concentrations in normoxia and hypoxia on the COL2 gene expression. The control (Day 0) samples were used as reference (set at 1). Ang-1 and Ang-2 were administrated in three concentrations: 10, 50 and 100 ng/mL. (**a**) COL2 relative gene expression of NP cells mixed population; (**b**) COL2 relative gene expression of NP cells Tie2$^-$ population; (**c**) COL2 relative gene expression of NP cells Tie2$^+$ population. Bars represent mean ± SD. A *p*-value < 0.05 was considered significant. Asterisks within figures denote the degree of statistical relevance observed: ** < 0.01; **** < 0.0001.

The *COL2* relative gene expression for the Ang-1/2 at 10 ng/mL and Ang-1 at 100 ng/mL sample in normoxia followed the same patterns concerning the peak values as described above (Figure 4b). The NP Tie2$^-$ population showed a similar relative gene expression, compared to the NP cells mixed population for the control day 7 sample and the Ang-2 at 100 ng/mL sample in normoxia and hypoxia. As showed previously in the NP cells mixed population, the effect of the stimulation of the 10 ng/mL concentration for Ang-1 and Ang-2 in normoxia is noteworthy. Additionally, the Ang-1 at 100 ng/mL sample reacted remarkably to its stimulation in normoxia and hypoxia. For all the other samples, no trend in the Ang-1 and Ang-2 administration was detectable.

Concerning the NP Tie2$^+$ population in normoxia, all the stimulated samples, as well as the control Day 7 sample, showed a 1.5- to 2.5-fold increase in their *COL2* relative gene expression, compared to the normoxia control Day 0 sample (Figure 4c). In hypoxia only, the control Day 7 sample with a 3.5-fold increase and the Ang-1/2 at 10 ng/mL sample with a 2- to 2.25-fold increase in their *COL2* relative gene expression showed a difference, compared to their control Day 0 sample. Interestingly, the 10 ng/mL concentration for Ang-1 and Ang-2 did not provoke a remarkable difference in the gene expression for normoxia as it did in the NP cells mixed population and the NP cells Tie2$^-$ population, compared to the normoxia control Day 0 sample. Summing up, the unstimulated control Day 7 sample in hypoxia reached the highest *COL2* gene expression in the NP cells Tie2$^+$ population.

3.4. Effect of Oxygen Tension and Angiopoietin-1/2 on Oxygen Level Related-Genes

Analysis of Hypoxia-Inducible Factor 1-Alpha Relative Gene Expression

The *HIF1A* relative gene expression in NP cells mixed population presented a similar trend, compared to the *COL2* relative gene expression in the NP cells mixed population. As in the *COL2* relative gene expression of NP cells mixed and NP Tie2$^-$ cells populations, the 10 ng/mL stimulation in normoxia of the Ang-1 and Ang-2 samples reached, also in this population, the highest amount of relative gene expression (Figure 5a). All stimulated hypoxia samples and the control Day 7 sample range in a 10% decrease and increase around the baseline in their *COL2* relative gene expression. Thus, we detected no effect in a dose-dependent manner of angiopoietin on the relative gene expression or specific variation in the use of Ang-1 or Ang-2 in hypoxia. However, the stimulation in normoxia clearly induced a higher potency than in hypoxia for 4 out of the 6 samples, whereas, again, no clear trend in concentration- or protein-dependency could be evaluated.

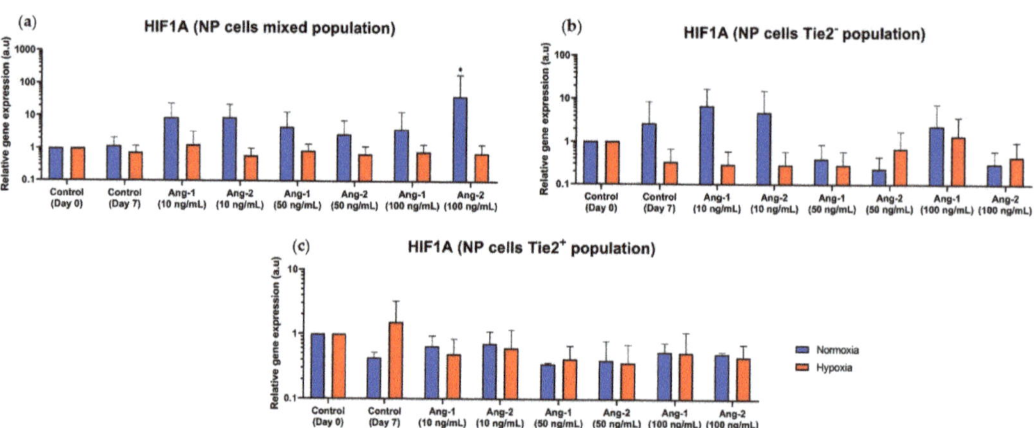

Figure 5. Relative gene expression of the three NP cells populations, comparing the effect of Ang1/2 at different concentrations in normoxia and hypoxia on the HIF1α gene expression. The control (Day 0) samples were used as reference (set at 1). Ang-1 and Ang-2 were administrated in three concentrations: 10, 50 and 100 ng/mL. (**a**) HIF1α relative gene expression of NP cells mixed population; (**b**) HIF1α relative gene expression of NP cells Tie2$^-$ population; (**c**) HIF1α relative gene expression of NP cells Tie2$^+$ population. Bars represent mean $\pm$ SD. A *p*-value < 0.05 was considered significant. Asterisks within figures denote the degree of statistical relevance observed: * < 0.05.

The normoxia associated samples of the Ang-1 and Ang-2 at 10 ng/mL concentrations reached a 12-fold and 15-fold increase in their *HIF1A* relative gene expression, compared to the control Day 0 sample (Figure 5b). Remarkably, similar to the NP cells mixed population, the NP Tie2$^-$ cells population showed a similar pattern in relative gene expression, compared to the *COL2* relative gene expression within the NP Tie2$^-$ cells population. It is also worth mentioning that the Ang-2 at 50 ng/mL and the Ang-1 at 100 ng/mL samples achieved the highest effect in stimulation in hypoxia for both *COL2* and *HIF1A* relative gene expression within the NP cells Tie2$^-$ population.

In the NP cells Tie2$^+$ population (Figure 5c), the incubation with Ang-1 and Ang-2 in their different concentrations does not seem to have any effect on the relative *HIF1A* gene expression. Notably, all of the stimulated samples showed downregulation in their gene expression, compared to the control conditions at Day 0.

3.5. Effect of Oxygen Tension and Angiopoietin-1/2 on NP Progenitor Cells Related-Genes

Analysis of Angiopoietin-1 Receptor Relative Gene Expression

In the NP cells mixed population, all stimulated samples in normoxia, as well as in hypoxia, showed an increase in their gene expression (Figure 6a). With a stimulation dose of

10 ng/mL for Ang-1/2, the gene expression in normoxia reached two peak values: the Ang-1 at 10 ng/mL sample presents a statistically relevant 400-fold increase and the Ang-2 at 10 ng/mL sample, a 200-fold upregulation of their *TEK* gene expression, compared to their control Day 0 sample. Thus, in this population, again, the 10 ng/mL dose in normoxia reached the highest values in gene expression. However, in hypoxia, no trend in gene expression upon stimulation by Ang-1 or Ang-2 and their different concentrations were detectable.

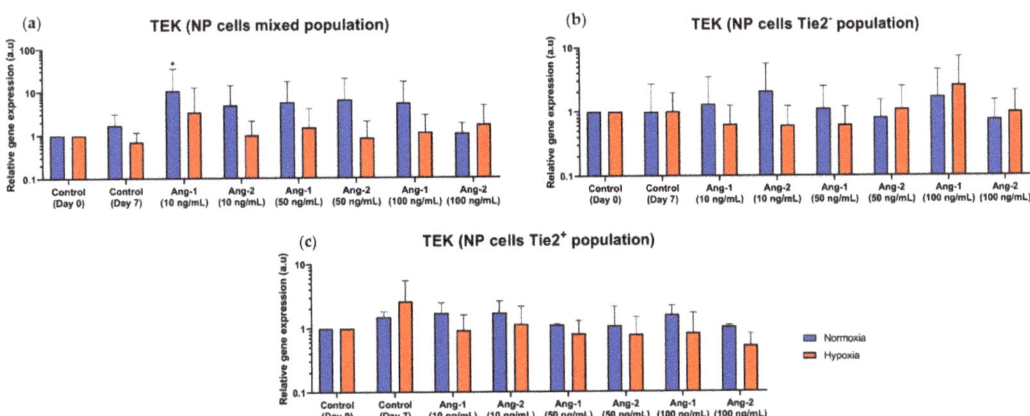

Figure 6. Relative gene expression of the three NP cells populations, comparing the effect of Ang-1/2 at different concentrations in normoxia and hypoxia on the TEK gene expression. The control (Day 0) samples were used as reference (set at 1). Ang-1 and Ang-2 were administrated in three concentrations: 10, 50 and 100 ng/mL. (**a**) TEK relative gene expression of NP cells mixed population; (**b**) TEK relative gene expression of NP cells Tie2$^-$ population; (**c**) TEK relative gene expression of NP cells Tie2$^+$ population. Bars represent mean ± SD. A p-value < 0.05 was considered significant. Asterisks within figures denote the degree of statistical relevance observed: * < 0.05.

Concerning the NP cells Tie2$^-$ population (Figure 6b), no clear trends in dose-dependency nor protein-dependency are detectable. In normoxia, for all the NP cells Tie2$^+$ samples treated with Ang-1 or Ang-2 and including the Day 7 control sample, the *TEK* relative gene expression was increased up to 2-fold, compared to the control Day 0 equivalent (Figure 6c). Therefore, normoxia upregulated the gene expression slightly. Yet, the stimulation with Ang-1 or Ang-2 in its different doses did not have any effect in normoxia, as the gene expression stays in a similar range for all normoxia samples. In addition, in the hypoxic condition, the stimulation did not have any upregulating effect. This missing reaction to stimulation in hypoxia was further confirmed by the control day 7 sample, which reached the maximum value of *TEK* relative gene expression for this population.

4. Discussion

Our study showed that the three different cells population studied here and present in the NP tissue act/behave differently in hypoxic and normoxic conditions. The increase in the O_2 tension significantly decreases the NP cells mixed population proliferation, while, conversely, we observed an increase in proliferation of the NP cells Tie2$^+$ population. Concerning the cell's activity, the decrease in the O_2 tension induced an increase in the cell's metabolism for the NP cells mixed and Tie2$^+$ populations.

Additionally, the three NP cells population showed variation in their phenotype/genotype, according to the presence of Ang-1 or Ang-2 in combination with normoxic or hypoxic conditions. To our knowledge, no other studies are comparing the influence of the oxygen tension combined on three distinct cells population, well-characterized in the human NP tissue, including NP progenitor cells. Moreover, to the best of our knowledge, as of today, there is no literature published about the different components of the heterogeneous NP tissue cell's population, including NPPCs isolated from relatively young and healthy IVDs in relation to the activation of the receptor-ligand system Tie2/Ang.

4.1. Ang-1/2 and Its Effect on DNA Measurement in Normoxic and Hypoxic Conditions

Evidence has arisen in recent years suggesting that a tissue renin–angiotensin system (tRAS) is involved in the progression of various human diseases [28], including IVD degeneration [29]. Numerous reports and studies showed the positive effects of pathologic tRAS pathway inhibition and protective tRAS pathway stimulation on the treatment of cardiovascular, inflammatory, and autoimmune disease and the progression of neuropathic pain. Cell proliferation, senescence, and aging are known to be related to RAS pathways. In this context, the DNA concentration measured in our study in all the NP cells mixed and Tie2^{+} population samples cultured within normoxic conditions were statistically different from the controls at Day 0. None of the samples in the NP cells Tie2^{-} population showed statistically significant differences. The amount of DNA in the NP cells mixed population decreased, whereas the amount was increased in the NP cells Tie2^{+} population on Day 7 for the control and the stimulated samples in normoxia, compared to their control Day 0 equivalents. In hypoxia, no effect on the proliferation was detected, except for the NP cells Tie2^{+} population treated with Ang-1 at 10 ng/mL.

In the NP cells Tie2^{-} population, a slight upregulation in the amount of DNA was seen in every sample, in normoxia and hypoxia, compared to the Day 0 sample. The reason for this discrepancy between the NP cells mixed and NP Tie2^{-} cells populations is not clear. As the NP cells mixed population consists of 95 to 99% of the exact cell types as the Tie2^{-} cells population, the 1 to 5% difference in cell type should not provoke the observed difference [1]. Furthermore, both samples underwent the same schedule of incubation and medium change, which should result in a more similar outcome. However, a possible cell-to-cell interaction or co-culture between the Tie2^{+} and Tie2^{-} cells within the mixed population could influence the proliferation as shown in other tissue and cell types [30,31].

All samples treated in the NP cells Tie2^{+} population showed an increased amount of DNA, compared to the Day 0 sample and a trend of higher amounts of DNA in normoxia, compared to hypoxia. A possible explanation for the lower amount of DNA detected within the cells population incubated under hypoxic conditions, compared to the normoxic condition, could be the potential differentiation within the chondrogenic lineage of the cells in hypoxia, reducing, at the same time, cell proliferation. As described by Lyu et al. [12], an explanation could be that hypoxia can exert either positive or negative effects on IVD progenitor cells. In particular, hypoxia can inhibit proliferation and induce apoptosis but might also promote the chondrogenic differentiation of IVD progenitor cells. Huang et al. [32] identified tissue-specific intervertebral disc progenitor cells (DPCs) from healthy Rhesus monkeys; in particular, they found that they are sensitive to low oxygen tension and undergo apoptosis under hypoxic conditions due to their inability to induce/stabilize hypoxia-inducible factors (HIF). These results confirm our findings. Our findings and the findings of the previously mentioned studies contradict the publication of Tekari et al. [13], who found a better cell survival of Tie2^{+} cells cultured in hypoxia (2%), which were based on young-aged bovine cells.

We identified in our results a trend for an increased amount of DNA in normoxia for the NP Tie2^{+} population. These are contradictory findings to several studies, in which it was shown that an oxemic shift would be expected to lead to a failure in progenitor cell activation [13,33]. The variation in oxygen tension is possibly mediated by alterations in the vascular supply to the endplate cartilage or even the annulus fibrosus. In turn, the cells no longer reside in their native associated microenvironment and are associated with a reduced cell proliferation capacity [34–36].

Concerning the amount of DNA, we detected either no effect in a dose-dependent manner of angiopoietin or specific variation in the use of Ang-1 or Ang-2 in normoxia and hypoxia for the three tested populations. Sakai et al. [1] demonstrated selective enhancement of the proliferation of Tie2^{+} GD2^{-} CD24^{-} (i.e., Tie2 single-cell sorted positive cells (T/sp)) and of Tie2$^+$ GD2$^+$ CD24$^-$ (i.e., Tie2 single-positive cells (TG/dp)) by using a co-culture with a cells feeder layer (murine stromal cells overexpressing human Ang-1 (AHESS-5). A marked Ang-1 dependent growth resulted in increased proliferation of the T/sp population by 3.1 $\pm$ 0.4 times and for the TG/dp population by 3.2 $\pm$ 0.4 times. Unfortunately, no information about the administrated Ang-1 dose or incubation time was published. Nevertheless, these results could not be confirmed with the NP cells Tie2$^+$ population in this study. A possible explanation could be that the Tie2 receptor was not expressed anymore over in vitro culture time, as cells needed expansion before starting the experiment. Summing up, at the time of incubation with Ang-1 and Ang-2, the cells were supposedly not Tie2$^+$ progenitor cells anymore, and therefore could not react to any stimulation, as the receptor for our ligands was not present anymore.

Ligand-deficient Ang1$^{-/-}$ or receptor-deficient Tie2$^{-/-}$ is embryonically lethal for transgenic mice. Both deletions showed common features in their phenotype, and both failed to develop a fully functional cardiovascular system [4]. Vasculogenesis proceeds normally, but remodeling and maturation of the vessels are defective [24]. Ang-2-deficient mice seem phenotypically normal at birth but die within 14 days or develop normally throughout adulthood. On the other hand, systemic embryonic Ang-2 overexpression results in an embryonic lethal phenotype of Ang-2 transgenic mice [18].

Furthermore, Wang et al. [37] determined the role of Ang-2 in the pathology of IVDD. They verified the expression of Ang-2 in human degenerative NPC and showed that Ang-2 expression was considerably higher in severe IVDD than in mild IVDD. Additionally, they demonstrated that NP cells treated with exogeneous Ang-2 resulted in marked downregulation of type II collagen and aggrecan and a significant increase in the expression of MMP13, ADAMTS4, and the pro-inflammatory cytokine IL-1β, both on protein and mRNA levels. The resulting dysregulation in ECM degradation in NPCs upon Ang-2 exposure contributed to the pathogenesis of IVDD. Conclusively, the study reveals Ang-2 inhibition as a potential novel therapeutic target for IVDD treatment.

4.2. Ang-1/2 and Its Effect on Metabolic Activity Per DNA in Normoxic & Hypoxic Conditions

Out of all analyses, in cell metabolism relative to DNA amount, the difference between oxygen tension ($\leq$ 5% vs. 21%) was in favor of hypoxia for every single sample tested in our three populations of interest, i.e., NP cells mixed, NP cells Tie2$^-$, and NP cells Tie2$^+$ populations. In our study, the samples in the hypoxic environment all achieved a higher metabolism, compared to the normoxic environment. This is in accordance with previous studies upon oxygen levels in NPC cultures, such as those of Feng et al. [38], Mwale et al. [39], Stoyanov et al. [40], and Gantenbein et al. [31]. The authors demonstrated that hypoxia enhances NPCs phenotype, which results in greater ECM components production and, indeed, a higher metabolism.

Additionally, as the NP cells mixed population and NP Tie2$^-$ cells population only differ in 1 to 5% of their cell population (only depletion of Tie2$^+$ cells), this could explain that stimulated samples showed comparable values in their metabolic activity [6]. In contrast, the NP cells Tie2$^+$ population responded in higher potency to the normoxic and hypoxic environments, compared to the two others population (i.e., mixed and Tie2$^-$). On one hand, the upregulation of the cell's metabolism would therefore seem logical, as this population expresses the Tie2 receptor and thus can react to the stimulation (i.e., presence/absence of Ang) by its ligands Ang-1 and Ang-2. However, on the other hand, the stimulation effect remains in question, as the treated samples do not show any trend either in a dose- or protein-dependent manner.

According to the findings of Wang et al. [23], Ang-2 plays a role in suppressing cell adhesion, decreasing cell viability, and promoting NP cell apoptosis. Concerning the cell's metabolism, in our study, the findings for Ang-2 could not be confirmed, which supports the aforementioned lack of effect upon Ang-1 and Ang-2 stimulation. Thus, the more reliable and convincing explanation for the upregulation of fluorescence intensity (i.e., cell's metabolism) in the NP cells Tie2$^+$ population in normoxia and hypoxia, compared to the NP cells mixed and NP cells Tie2^{+} populations, could potentially be that progenitor cells show a higher metabolism, compared to mature cells [41].

Another aspect that has to be taken into consideration by comparing the findings of Wang et al. [23] with our study is that the cells from our study were isolated from trauma discs and not from degenerated IVDs. Thus, the cells in Wang's study were already stimulated by Ang-1, due to potential blood vessel infiltration and inflammation in the degenerated disc as well as by Ang-2, as the Ang-2/Ang-1 ratio increases in degenerated discs for the same reason [37]. Whether such contradictory findings could be related to differences in IVD degeneration pathways between injuries (trauma), natural aging, or the degree of IVD degeneration is still unclear.

4.3. *Ang-1/2 and Its Effect on the ECM Gene Expression in Normoxia & Hypoxia*

Looking at the gene expression levels, we did not observe significant results between normoxia and hypoxia for the majority of the samples. The effect of hypoxia on gene expression associated with chondrogenic lineage and ECM production in the IVD is already well documented in the literature [38,39]. At least, in the NP cell Tie2^{+} population for ACAN, we detected a trend of upregulation in hypoxia, whereas for *COL2*, it was not the case. If Ang-1 and/or Ang-2 had a stimulatory effect, this would be the case for the NP cells Tie2^{+} population, as this population has the Tie2 receptor for the binding of the Ang-ligands. It is, therefore, astonishing and unexpected to find the highest upregulations for ACAN in normoxia in the NP cells mixed and Tie2^{-} populations for Ang-1/2 at 10 ng/mL and Ang-1 at 100 ng/mL. A possible explanation for these particular increases could be related to any cross-reactions of the ligands Ang-1 and Ang-2 with another receptor. This hypothesis is difficult to defend in the sense that there are only two or three out of the six treated samples in normoxia that show an effect upon stimulation.

However, the main explanation for the stimulation of Ang-1 and Ang-2 not having any effect is the outcome for the NP cells Tie2^{+} population. The stimulation did not show any differences upon Ang-1 or Ang-2 protein stimulation, including their different concentrations. In the case of the NP cells Tie2^{+} population, several explanations can be considered. First, there could have been a cross-reaction of the Tie2 receptor with other ligands, which subsequently blocked the binding of Ang-1/2 to the receptor and thus the downstream pathways. Second, it would be possible that the Tie2 receptor was not expressed after the expansion culture phase. Briefly, at the time of incubation with Ang-1 and Ang-2, the cells were not Tie2^{+} progenitor cells anymore. To confirm the progenitor state of the cell right before the start of the experiment, another FACS analysis could have been performed. Third, the first FACS analysis itself could already have stressed the cells and favored the loss of their progenitor attributes.

The hypothesis for the lack of effect of Ang stimulation is encouraged by the examination of the TEK gene. It is eminently important to see that, similar to *ACAN* and *COL2*, we did miss out on any specific reaction in the relative gene expression of *TEK* to either Ang-1, Ang-2, including their three different doses in the Tie2^{+} cells population. This is unexpected for the same reasons as already mentioned for *ACAN* and *COL2*. In the NP mixed and NP Tie2^{-} cells populations, the relative gene expression in *TEK* and *HIF1A* do show the same patterns as *COL2* and *ACAN*. The peak values in normoxia deserve special mention, such as Ang-1/2 at 10 ng/mL and Ang 1 at 100 ng/mL. However, again, it was unexpected to find such upregulation in normoxia of the genes within the NP mixed and NP cells Tie2^{-} populations. As we find the same pattern several times, there must be attractive or favorable attributes for these doses in normoxia. Up to that point, explanations remain elusive. As already mentioned before, the data must be interpreted with caution, due to the high donor variation. The downregulation of *HIF1A* in hypoxia in the NP cells mixed, NP cells Tie2^{-} and Tie2^{+} populations can be explained, as hypoxia progressively decreases *HIF1A* expression at the mRNA level, due to the mRNA-destabilizing protein tristetraprolin [42].

However, Huang et al. [32] detected that the level of relative gene expression of *HIF1A* for DPCs of the NP remained constant under hypoxia, compared to normoxia. They suggest that incompetency of the DPCs of the NP to adapt to oxygen tension may be in part due to the inability to upregulate HIF expression. Their findings are consistent with our results of the *HIF1A* relative gene expression of the NP cells Tie2$^+$ population.

4.4. Study Weaknesses and Limitations

One of the main limitations of this study was the relatively small sample size for the NP cells Tie2$^+$ population. Sample expansion of the NP progenitor cells after a sorting process was a challenge, due to the relatively low percentage who represent this population in the NP tissue, and future studies should recruit IVD material from bigger cohorts and multiple populations and etiologies. Moreover, previous studies demonstrated the difficulties to expand, in vitro, this population [16,43]. Additionally, the "controls" used in the study were not truly healthy IVDs, as they were traumatic discs. In this respect, a comparison of fresh cadaveric IVD material of undegenerated nature would be essential for any future investigations. Moreover, the 2D culture was shown to not be the most accurate system for culture and expansion of NP progenitor cells [16]. Previous studies, such as that of Sakai et al., dealt with only Tie2$^+$ cells from degenerated discs; Wang et al. investigated the effect of Ang-2 in degenerated discs. It remains unclear whether the Tie2$^+$ cells from the degenerated or the trauma discs react in a comparable way upon Ang stimulation. Thus, it remains unclear whether we can compare our results (cells isolated from trauma patients) to the results from publications based on degenerated discs. We suspect that the results from degenerated discs are potentially biased, as discs subjected to degeneration present a higher Ang-2/Ang-1 ratio than healthy discs [23], and the Ang-1 release should be increased as the degenerated discs present blood vessels ingrowth.

Another strong limitation of the study is that we only used the Tie2$^-$ population as a control population. Previous studies [1,37] investigated Tie2 blocking antibodies as a negative control. For our study, a second negative control with a Tie2 blocking antibody would have made sense, as we observed, especially in the qPCR analysis, the effects in normoxia for Ang-1/2 at 10 ng/mL and Ang-1 at 100 ng/mL. Alternatively, other small inhibitory molecules could have been tested, which are known to interfere and block the Ang-1 and/or Ang-2 receptors. These inhibitors could have been tested by adding these in equimolar concentrations or by titration approaches. These are important negative controls that were not considered here.

4.5. Outlook and Clinical Relevance

In this study, we used human NP cells isolated from trauma patients' IVDs. We demonstrated the significant impact of culturing NP cells under hypoxic conditions to improve the proliferation, metabolism and gene expression of NP progenitor cells. However, no impact or effects were observed concerning the implementation of Ang-1 or Ang-2 in our culture medium.

The ensuing findings could directly impact clinical and therapeutic strategies in contrast to preclinical studies in other species that often need to be verified in humans first. Additionally, we provided novel evidence that culture within a hypoxic environment can drastically influence the proliferation, metabolism, and gene expression of the different cell populations present in human NP tissue. Moreover, we demonstrated that the hypoxic conditions affect more the progenitor cell populations from the human NP tissue (also known as Tie2$^+$) than the two other populations (mixed and Tie2$^-$). All those findings could lead to a new way to culture NP progenitors cells prior to clinical applications. Several studies have demonstrated Ang 1 and Ang 2 act in a concentration-dependent manner [17,44–47]. Beyond the crucial role in blood vessel development, recent studies suggest a broader role for Tie2, and the importance of Ang-1 and Ang-2 seem to be evident, as they are ligands (agonists) of the Tie2$^+$ receptor and, therefore, expressed in NPPC. Nevertheless, previous studies, such as those of Sakai et al. and Wang et al., demonstrated the effects of Ang-1 on cell proliferation and of Ang-2 on cell viability and cell adhesion. Their findings prompt us to clarify the favorable methods, adjust and improve our study design, and to further investigate the effects of Ang-1 and Ang-2 on NP cells. Despite our results, Ang-1 and Ang-2 may constitute a novel therapeutic target for the treatment of IVDD. A better understanding of the influence of Ang-1 and Ang-2 on the NPPCs would be of eminent importance in the context of IVD tissue engineering and regenerative medicine.

Other parameters that could have been taken into consideration as well are the GAG content, and the measurement of inflammatory cytokines. Moreover, HIF1A should be investigated at the protein level with Western blot analysis or immunohistology, as the HIF1A mRNA expression could have a fast turnover, especially in prolonged hypoxia [42].

5. Conclusions

- To the best of our knowledge, this study cannot provide evidence for stimulation of Ang-1 or Ang-2 or their dose-dependent administration influences on NP cell proliferation.
- Neither on the NP cell's metabolism, or relative gene expression.
- However, our study demonstrated that hypoxia enhances NP cells phenotype, which resulted in greater ECM components production, in accordance with Feng et al. [38] and Mwale et al. [39].
- Moreover, we also demonstrated a higher NP cell metabolism if cultured in hypoxia.
- Regarding the differences in Tie2$^+$ cell proliferation upon normoxia and hypoxia, our results confirmed the studies by Lyu et al. [12] and Huang et al. [32].

Author Contributions: Conceptualization, M.C.B., B.G. and J.G.; methodology, M.C.B. and A.O.; formal analysis, M.C.B. and J.G.; investigation, M.C.B. and J.G.; resources, C.E.A. and B.G.; data curation, M.C.B. and J.G.; writing—original draft preparation, M.C.B. and J.G.; writing—review and editing, M.C.B., B.G., A.S.C., S.H., K.A.C.O. and J.G.; visualization, J.G. and B.G.; supervision, J.G. and B.G.; project administration, J.G. and B.G.; funding acquisition, B.G. and C.E.A. All authors have read and agreed to the published version of the manuscript.

Funding: Financial support was received from the iPSpine H2020 project #825925 entrusted to B.G, www.iPSpine.eu (accessed on 1 August 2021), https://cordis.europa.eu/project/id/825925 (accessed on 1 August 2021). Further funds were received from the Swiss Society of Orthopedics (SGOT), the clinical trials unit of Bern University Hospital, and by a Eurospine Task Force Research grant #2019_22 entrusted to C.E.A.

Institutional Review Board Statement: The study was conducted according to the guidelines of the Declaration of Helsinki and approved by the Ethics Committee of University of Bern (# 2019-00097).

Informed Consent Statement: Informed consent was obtained from all subjects involved in the study.

Data Availability Statement: The data presented in this study are available on request from the corresponding author.

Acknowledgments: The cytometer equipment was from FACSlab core facility of the University of Bern (Available online: www.facslab.unibe.ch (accessed on 1 August 2021).

Conflicts of Interest: The authors declare no conflict of interest.

References

1. Sakai, D.; Nakamura, Y.; Nakai, T.; Mishima, T.; Kato, S.; Grad, S.; Alini, M.; Risbud, M.V.; Chan, D.; Cheah, K.S.; et al. Exhaustion of nucleus pulposus progenitor cells with ageing and degeneration of the intervertebral disc. *Nat. Commun.* **2012**, *3*, 1264. [CrossRef] [PubMed]
2. Baumgartner, L.; Wuertz-Kozak, K.; Le Maitre, C.L.; Wignall, F.; Richardson, S.M.; Hoyland, J.; Ruiz Wills, C.; Gonzalez Ballester, M.A.; Neidlin, M.; Alexopoulos, L.G.; et al. Multiscale regulation of the intervertebral disc: Achievements in experimental, in silico, and regenerative research. *Int. J. Mol. Sci.* **2021**, *22*, 703. [CrossRef] [PubMed]
3. Roberts, S.; Evans, E.H.; Kletsas, D.; Jaffray, D.C.; Eisenstein, S.M. Senescence in human intervertebral discs. *Eur. Spine J.* **2006**, *15* (Suppl. 3), S312–S316. [CrossRef] [PubMed]
4. Adams, M.A.; Roughley, P.J. What is intervertebral disc degeneration, and what causes it? *Spine* **2006**, *31*, 2151–2161. [CrossRef] [PubMed]

5. Binch, A.L.A.; Fitzgerald, J.C.; Growney, E.A.; Barry, F. Cell-based strategies for ivd repair: Clinical progress and translational obstacles. *Nat. Rev. Rheumatol.* **2021**, *17*, 158–175. [CrossRef] [PubMed]
6. Frauchiger, D.A.; Tekari, A.; May, R.D.; Dzafo, E.; Chan, S.C.W.; Stoyanov, J.; Bertolo, A.; Zhang, X.; Guerrero, J.; Sakai, D.; et al. Fluorescence-activated cell sorting is more potent to fish intervertebral disk progenitor cells than magnetic and beads-based methods. *Tissue Eng. Part C Methods* **2019**, *25*, 571–580. [CrossRef] [PubMed]
7. Kumar, D.; Gerges, I.; Tamplenizza, M.; Lenardi, C.; Forsyth, N.R.; Liu, Y. Three-dimensional hypoxic culture of human mesenchymal stem cells encapsulated in a photocurable, biodegradable polymer hydrogel: A potential injectable cellular product for nucleus pulposus regeneration. *Acta Biomater.* **2014**, *10*, 3463–3474. [CrossRef]
8. Urban, J.P.; Smith, S.; Fairbank, J.C. Nutrition of the intervertebral disc. *Spine* **2004**, *29*, 2700–2709. [CrossRef]
9. Kwon, W.K.; Moon, H.J.; Kwon, T.H.; Park, Y.K.; Kim, J.H. The role of hypoxia in angiogenesis and extracellular matrix regulation of intervertebral disc cells during inflammatory reactions. *Neurosurgery* **2017**, *81*, 867–875. [CrossRef] [PubMed]
10. Bartels, E.M.; Fairbank, J.C.; Winlove, C.P.; Urban, J.P. Oxygen and lactate concentrations measured in vivo in the intervertebral discs of patients with scoliosis and back pain. *Spine* **1998**, *23*, 1–7, discussion 8. [CrossRef]
11. Tejada, S.; Batle, J.M.; Ferrer, M.D.; Busquets-Cortes, C.; Monserrat-Mesquida, M.; Nabavi, S.M.; Del Mar Bibiloni, M.; Pons, A.; Sureda, A. Therapeutic effects of hyperbaric oxygen in the process of wound healing. *Curr. Pharm. Des.* **2019**, *25*, 1682–1693. [CrossRef] [PubMed]
12. Lyu, F.J.; Cheung, K.M.; Zheng, Z.; Wang, H.; Sakai, D.; Leung, V.Y. Ivd progenitor cells: A new horizon for understanding disc homeostasis and repair. *Nat. Rev. Rheumatol.* **2019**, *15*, 102–112. [CrossRef]
13. Tekari, A.; Chan, S.C.W.; Sakai, D.; Grad, S.; Gantenbein, B. Angiopoietin-1 receptor tie2 distinguishes multipotent differentiation capability in bovine coccygeal nucleus pulposus cells. *Stem Cell Res. Ther.* **2016**, *7*, 75. [CrossRef] [PubMed]
14. Sakai, D.; Schol, J. Cell therapy for intervertebral disc repair: Clinical perspective. *J. Orthop. Translat.* **2017**, *9*, 8–18. [CrossRef] [PubMed]
15. Sakai, D.; Schol, J.; Bach, F.C.; Tekari, A.; Sagawa, N.; Nakamura, Y.; Chan, S.C.W.; Nakai, T.; Creemers, L.B.; Frauchiger, D.A.; et al. Successful fishing for nucleus pulposus progenitor cells of the intervertebral disc across species. *JOR Spine* **2018**, *1*, e1018. [CrossRef] [PubMed]
16. Guerrero, J.; Hackel, S.; Croft, A.S.; Albers, C.E.; Gantenbein, B. The effects of 3d culture on the expansion and maintenance of nucleus pulposus progenitor cell multipotency. *JOR Spine* **2021**, *4*, e1131. [CrossRef] [PubMed]
17. Bogdanovic, E.; Nguyen, V.P.; Dumont, D.J. Activation of tie2 by angiopoietin-1 and angiopoietin-2 results in their release and receptor internalization. *J. Cell Sci.* **2006**, *119*, 3551–3560. [CrossRef] [PubMed]
18. Fiedler, U.; Augustin, H.G. Angiopoietins: A link between angiogenesis and inflammation. *Trends Immunol.* **2006**, *27*, 552–558. [CrossRef] [PubMed]
19. Thurston, G.; Daly, C. The complex role of angiopoietin-2 in the angiopoietin-tie signaling pathway. *Cold Spring Harb. Perspect. Med.* **2012**, *2*, a006550. [CrossRef] [PubMed]
20. Wakui, S.; Yokoo, K.; Muto, T.; Suzuki, Y.; Takahashi, H.; Furusato, M.; Hano, H.; Endou, H.; Kanai, Y. Localization of ang-1, -2, tie-2, and vegf expression at endothelial-pericyte interdigitation in rat angiogenesis. *Lab. Investig.* **2006**, *86*, 1172–1184. [CrossRef] [PubMed]
21. Akwii, R.G.; Sajib, M.S.; Zahra, F.T.; Mikelis, C.M. Role of angiopoietin-2 in vascular physiology and pathophysiology. *Cells* **2019**, *8*, 471. [CrossRef] [PubMed]
22. Imanishi, Y.; Hu, B.; Jarzynka, M.J.; Guo, P.; Elishaev, E.; Bar-Joseph, I.; Cheng, S.Y. Angiopoietin-2 stimulates breast cancer metastasis through the alpha(5)beta(1) integrin-mediated pathway. *Cancer Res.* **2007**, *67*, 4254–4263. [CrossRef] [PubMed]
23. Wang, K.; Liu, W.; Song, Y.; Wu, X.; Zhang, Y.; Li, S.; Gao, Y.; Tu, J.; Liu, Y.; Yang, C. The role of angiopoietin-2 in nucleus pulposus cells during human intervertebral disc degeneration. *Lab. Investig.* **2017**, *97*, 971–982. [CrossRef] [PubMed]
24. Eklund, L.; Olsen, B.R. Tie receptors and their angiopoietin ligands are context-dependent regulators of vascular remodeling. *Exp. Cell Res.* **2006**, *312*, 630–641. [CrossRef] [PubMed]
25. Makinde, T.O.; Agrawal, D.K. Increased expression of angiopoietins and tie2 in the lungs of chronic asthmatic mice. *Am. J. Respir Cell Mol. Biol* **2011**, *44*, 384–393. [CrossRef] [PubMed]
26. McCarter, S.D.; Mei, S.H.; Lai, P.F.; Zhang, Q.W.; Parker, C.H.; Suen, R.S.; Hood, R.D.; Zhao, Y.D.; Deng, Y.; Han, R.N.; et al. Cell-based angiopoietin-1 gene therapy for acute lung injury. *Am. J. Respir. Crit. Care Med.* **2007**, *175*, 1014–1026. [CrossRef] [PubMed]
27. Livak, K.J.; Schmittgen, T.D. Analysis of relative gene expression data using real-time quantitative pcr and the 2(-delta delta c(t)) method. *Methods* **2001**, *25*, 402–408. [CrossRef] [PubMed]
28. Saravi, B.; Li, Z.; Lang, C.N.; Schmid, B.; Lang, F.K.; Grad, S.; Alini, M.; Richards, R.G.; Schmal, H.; Sudkamp, N.; et al. The tissue renin-angiotensin system and its role in the pathogenesis of major human diseases: Quo vadis? *Cells* **2021**, *10*, 650. [CrossRef] [PubMed]
29. Li, Z.; Wystrach, L.; Bernstein, A.; Grad, S.; Alini, M.; Richards, R.G.; Kubosch, D.; Sudkamp, N.; Izadpanah, K.; Kubosch, E.J.; et al. The tissue-renin-angiotensin-system of the human intervertebral disc. *Eur. Cell Mater.* **2020**, *40*, 115–132. [CrossRef] [PubMed]
30. Bertolo, A.; Thiede, T.; Aebli, N.; Baur, M.; Ferguson, S.J.; Stoyanov, J.V. Human mesenchymal stem cell co-culture modulates the immunological properties of human intervertebral disc tissue fragments in vitro. *Eur. Spine J.* **2011**, *20*, 592–603. [CrossRef] [PubMed]

31. Gantenbein, B.; Calandriello, E.; Wuertz-Kozak, K.; Benneker, L.M.; Keel, M.J.; Chan, S.C. Activation of intervertebral disc cells by co-culture with notochordal cells, conditioned medium and hypoxia. *BMC Musculoskelet. Disord.* **2014**, *15*, 422. [CrossRef] [PubMed]
32. Huang, S.; Leung, V.Y.; Long, D.; Chan, D.; Lu, W.W.; Cheung, K.M.; Zhou, G. Coupling of small leucine-rich proteoglycans to hypoxic survival of a progenitor cell-like subpopulation in rhesus macaque intervertebral disc. *Biomaterials* **2013**, *34*, 6548–6558. [CrossRef] [PubMed]
33. Risbud, M.V.; Shapiro, I.M. Notochordal cells in the adult intervertebral disc: New perspective on an old question. *Crit. Rev. Eukaryot. Gene Expr.* **2011**, *21*, 29–41. [CrossRef] [PubMed]
34. Yasen, M.; Fei, Q.; Hutton, W.C.; Zhang, J.; Dong, J.; Jiang, X.; Zhang, F. Changes of number of cells expressing proliferation and progenitor cell markers with age in rabbit intervertebral discs. *Acta Biochim. Biophys. Sin.* **2013**, *45*, 368–376. [CrossRef] [PubMed]
35. Jia, Z.; Yang, P.; Wu, Y.; Tang, Y.; Zhao, Y.; Wu, J.; Wang, D.; He, Q.; Ruan, D. Comparison of biological characteristics of nucleus pulposus mesenchymal stem cells derived from non-degenerative and degenerative human nucleus pulposus. *Exp. Ther. Med.* **2017**, *13*, 3574–3580. [CrossRef] [PubMed]
36. Zhao, Y.; Jia, Z.; Huang, S.; Wu, Y.; Liu, L.; Lin, L.; Wang, D.; He, Q.; Ruan, D. Age-related changes in nucleus pulposus mesenchymal stem cells: An in vitro study in rats. *Stem Cells Int.* **2017**, *2017*, 6761572. [CrossRef]
37. Wang, K.; Kang, L.; Liu, W.; Song, Y.; Wu, X.; Zhang, Y.; Hua, W.; Zhao, K.; Li, S.; Tu, J.; et al. Angiopoietin-2 promotes extracellular matrix degradation in human degenerative nucleus pulposus cells. *Int. J. Mol. Med.* **2018**, *41*, 3551–3558. [CrossRef] [PubMed]
38. Feng, G.; Li, L.; Liu, H.; Song, Y.; Huang, F.; Tu, C.; Shen, B.; Gong, Q.; Li, T.; Liu, L.; et al. Hypoxia differentially regulates human nucleus pulposus and annulus fibrosus cell extracellular matrix production in 3d scaffolds. *Osteoarthr. Cartil.* **2013**, *21*, 582–588. [CrossRef] [PubMed]
39. Mwale, F.; Ciobanu, I.; Giannitsios, D.; Roughley, P.; Steffen, T.; Antoniou, J. Effect of oxygen levels on proteoglycan synthesis by intervertebral disc cells. *Spine* **2011**, *36*, E131–E138. [CrossRef]
40. Stoyanov, J.V.; Gantenbein-Ritter, B.; Bertolo, A.; Aebli, N.; Baur, M.; Alini, M.; Grad, S. Role of hypoxia and growth and differentiation factor-5 on differentiation of human mesenchymal stem cells towards intervertebral nucleus pulposus-like cells. *Eur. Cell Mater.* **2011**, *21*, 533–547. [CrossRef]
41. Guehring, T.; Wilde, G.; Sumner, M.; Grunhagen, T.; Karney, G.B.; Tirlapur, U.K.; Urban, J.P. Notochordal intervertebral disc cells: Sensitivity to nutrient deprivation. *Arthritis Rheum.* **2009**, *60*, 1026–1034. [CrossRef] [PubMed]
42. Chamboredon, S.; Ciais, D.; Desroches-Castan, A.; Savi, P.; Bono, F.; Feige, J.J.; Cherradi, N. Hypoxia-inducible factor-1alpha mrna: A new target for destabilization by tristetraprolin in endothelial cells. *Mol. Biol. Cell* **2011**, *22*, 3366–3378. [CrossRef] [PubMed]
43. Zhang, X.; Guerrero, J.; Croft, A.S.; Albers, C.E.; Hackel, S.; Gantenbein, B. Spheroid-like cultures for expanding angiopoietin receptor-1 (aka. Tie2) positive cells from the human intervertebral disc. *Int. J. Mol. Sci.* **2020**, *21*, 9423. [CrossRef] [PubMed]
44. Kanda, S.; Miyata, Y.; Mochizuki, Y.; Matsuyama, T.; Kanetake, H. Angiopoietin 1 is mitogenic for cultured endothelial cells. *Cancer Res.* **2005**, *65*, 6820–6827. [CrossRef] [PubMed]
45. Hashimoto, T.; Wu, Y.; Boudreau, N.; Li, J.; Matsumoto, M.; Young, W. Regulation of tie2 expression by angiopoietin–potential feedback system. *Endothelium* **2004**, *11*, 207–210. [CrossRef] [PubMed]
46. Kim, I.; Kim, J.H.; Moon, S.O.; Kwak, H.J.; Kim, N.G.; Koh, G.Y. Angiopoietin-2 at high concentration can enhance endothelial cell survival through the phosphatidylinositol 3′-kinase/akt signal transduction pathway. *Oncogene* **2000**, *19*, 4549–4552. [CrossRef] [PubMed]
47. Kim, I.; Kim, H.G.; So, J.N.; Kim, J.H.; Kwak, H.J.; Koh, G.Y. Angiopoietin-1 regulates endothelial cell survival through the phosphatidylinositol 3′-kinase/akt signal transduction pathway. *Circ. Res.* **2000**, *86*, 24–29. [CrossRef] [PubMed]

Article

EGR2, IGF1 and IL6 Expression Are Elevated in the Intervertebral Disc of Patients Suffering from Diffuse Idiopathic Skeletal Hyperostosis (DISH) Compared to Degenerative or Trauma Discs

Benjamin Gantenbein [1,2,*,†], Rahel D. May [1,†], Paola Bermudez-Lekerika [1,2], Katharina A. C. Oswald [2], Lorin M. Benneker [2] and Christoph E. Albers [2]

[1] Tissue Engineering for Orthopaedics and Mechanobiology, Department for BioMedical Research (DBMR), University of Bern, CH-3008 Bern, Switzerland; Rayel.May@gmail.com (R.D.M.); paola.bermudez@dbmr.unibe.ch (P.B.-L.)

[2] Department of Orthopaedic Surgery and Traumatology, Inselspital, University of Bern, CH-3010 Bern, Switzerland; katharina.oswald@insel.ch (K.A.C.O.); lorin.benneker@insel.ch (L.M.B.); christoph.albers@insel.ch (C.E.A.)

* Correspondence: benjamin.gantenbein@dbmr.unibe.ch or benjamin.gantenbein@insel.ch; Tel.: +41-31-632-88-15

† These authors contributed equally.

Abstract: Diffuse idiopathic skeletal hyperostosis (DISH) is characterised by ectopic ossification along the anterior spine and the outer intervertebral discs (IVD). However, the centre of the IVD, i.e., the nucleus pulposus, always remains unaffected, which could be due to the inhibition of the bone morphogenetic protein (BMP) pathway. In this study, we investigated the transcriptome for the BMP pathway of DISH-IVD cells versus disc cells of traumatic or degenerative discs. The disc cells originated from nucleus pulposus (NP), annulus fibrosus (AF) and from cartilaginous endplate (CEP) tissue. Here, ninety genes of the transforming growth factor β-BMP signalling pathway were screened by qPCR. Furthermore, the protein expression of genes of interest was further investigated by immune-staining and semi-quantitative microscopy. IVDs of three DISH patients were tested against three control patients (same disc level and similar age). Early Growth Response 2 (*EGR2*) and Interleukin 6 (*IL6*) were both significantly up-regulated in DISH-IVD cells compared to controls (12.8 ± 7.6-fold and 54.0 ± 46.5-fold, respectively, means ± SEM). Furthermore, Insulin-like Growth Factor 1 (IGF1) tended to be up-regulated in DISH-IVD donors, i.e., 174.13 ± 120.6-fold. IGF1 was already known as a serum marker for DISH and other rheumatoid diseases and is confirmed here to play a possible key role in DISH-IVD.

Keywords: low back pain; diffuse idiopathic skeletal hyperostosis; RNA extraction; qPCR; TGFβ-pathway; immune-histochemistry; Interleukin 6; Insulin-like growth factor 1; early growth response 2

Citation: Gantenbein, B.; May, R.D.; Bermudez-Lekerika, P.; Oswald, K.A.C.; Benneker, L.M.; Albers, C.E. EGR2, IGF1 and IL6 Expression Are Elevated in the Intervertebral Disc of Patients Suffering from Diffuse Idiopathic Skeletal Hyperostosis (DISH) Compared to Degenerative or Trauma Discs. *Appl. Sci.* **2021**, *11*, 4072. https://doi.org/10.3390/app11094072

Academic Editor: Roger Narayan

Received: 26 February 2021
Accepted: 25 April 2021
Published: 29 April 2021

1. Introduction

Diffuse idiopathic skeletal hyperostosis (DISH), also known as Forestier's disease, was first described by Forestier and Rotes-Querol in 1950 [1]. The authors characterised this disorder by spinal stiffness, osteophytosis, and aberrant new-bone formation along the spine's anterior lateral aspect. The underlying causes are yet unknown and thus are called "idiopathic" [2]. DISH is generally characterised by ossification of ligamentous insertions (fibro-osteosis) and ossification or calcifications of tendons, ligaments, fasciae, joint capsule and annulus fibrosus (AF) fibres [3–5]. Furthermore, DISH increases the amount of normal cancellous and cortical bone. Moreover, a tendency of post-operative heterotopic bone formation has been observed [6]. The incidence of DISH, however, remains unknown, but it seems that factors of metabolic, endocrine, genetic, and epigenetics seem to be important in the development of this condition [4,7]. This disease mainly affects the elderly and is more frequent in men (65%) than in women, and the prevalence also increases with age (48–85) [8,9]. DISH patients often suffer from severe pain and need special treatment. Hence, the disease is often co-diagnosed with osteoarthritis [10,11].

Recently, Kuperus et al. (2016) [12] investigated the relationship between DISH and IVD degeneration. These authors found that partial or complete bone bridges were often present in DISH cadaveric IVDs that were only rarely found in healthy cadaveric controls. However, the disc height and degeneration level of DISH and control specimens were comparable. This specific feature also distinguishes the DISH condition from spondylosis [11].

To date, only very few molecular pathway studies have been conducted on the effect of DISH on IVDs. ElMiedany et al. (2000) proposed that the new formation of bone is mainly driven by abnormal cell growth/activity in the bony-ligamentous region [13]. Previous studies on the metabolomics of this disease found that patients suffering from symptomatic DISH disease have elevated levels of growth hormone (GH) and insulin-like growth factor 1 (IGF1) compared to patients of the same age and gender [9,14–19]. Thus, the molecular mechanism of DISH and the specific characteristics of underlying cell signalling remain obscure. Bone morphogenic proteins (BMPs), such as BMP2, BMP4, BMP6, BMP7, and BMP9, have shown their potential to induce the osteogenic differentiation of stromal cells and osteoblastic lineage cells in vitro and in vivo [20–23]. Furthermore, the recent literature has summarised and questioned the supra-physiological usage of BMP2 for the application of improved spinal fusion [24]. Thus, recently it has been hypothesised that BMP inhibitors could be the main reason for the lacking effect of BMP2 for improved spinal fusion [25–28]. In this context, it is of uttermost interest to analyse the transcriptome of DISH-IVDs in more detail, as all the surrounding tissue starts to ossify but not the IVD itself, which seems to be resistant to osteogenic signalling [29].

Hence, the current study aimed to compare the gene expression at the BMP pathway of cells of IVDs from patients suffering from DISH disease and compare it with cells originating from traumatic or degenerative IVDs. We assumed that trauma discs, especially from younger patients, contain presumably "healthier" cells that could be defined as a "control". Here, we investigated whether DISH disease is linked to alterations in BMP/TGFβ signalling pathways to gain further insights into how the DISH-IVD was affected by this idiopathic disease at the molecular level. We hypothesised that IVDs of patients diagnosed with DISH demonstrate specific gene expression changes affecting the TGFβ/BMP pathway compared to degenerative IVDs or trauma patients.

2. Materials and Methods

2.1. IVD Donor Materials and Cell Isolation

Human IVDs were obtained from patients suffering from DISH disease. As controls, traumatic or degenerative IVDs of patients undergoing spinal surgery were used (Table 1). Patients gave their written consent, and procedures were approved by the ethics committee of the Canton of Bern. Tissues were processed within 24 h after surgery.

Human IVD tissue was separated into nucleus pulposus (NP), annulus fibrosus (AF) and cartilaginous endplate (CEP) tissue by an experienced surgeon or processed as mixed IVD. Cell isolation from the native extracellular matrix was performed as described earlier [28]. Briefly, tissue was enzymatically digested by 1 h in pronase (Roche, Basel, Switzerland) followed by overnight incubation in collagenase type 2 (Worthington, London, United Kingdom). Subsequent expansion of NP cells (NPC), AF cells (AFC), and CEP cells (CEPC) followed in proliferation medium (low-glucose (1g/L) Dulbecco's Modified Eagle Medium (LG-DMEM) supplemented with 10% FBS and 1% P/S).

IVD cells were obtained from six DISH donors aged from 65 to 84 years (72.3 $\pm$ 3.1 years) and three "control" donors, which were lacking symptoms of DISH, suffering from degenerated or traumatic discs aged from 17 to 66 years (45 $\pm$ 14.5 years).

Table 1. Donor list of intervertebral disc cells. Cell type:nucleus pulposus (N), annulus fibrosus (A), cartilaginous endplate (CEP), C = cervical, L = lumbar, intervertebral disc mixed (IVD) Sex: Female (F), Male (M). Disc condition: Degenerated (D), Trauma (T), Pfirrmann Grade (PG).

Donor Groups Used in this Study										
DISH					**Control**					
Patient #	Cell Type	Age	Sex	Spinal Level	Patient #	Cell Type	Age	Sex	Spinal Level	Disc Condition
1	N/A/CEP	72	M	L4/L5	1	N/A/CE	66	F	L4/L5	D; PG3
2	N/A/CEP	65	F	L11/T12	2	N/A/CE	52	M	L11/T12	T; PG3
3	IVD	66	M	L5/L6	3	IVD	17	M	C5/C6	T
4	N/A	68	M	L11/T12	2	N/A/CE	52	M	L11/T12	T; PG3
5	A/CEP	79	M	C5/C6	3	IVD	17	M	C5/C6	T
6	A	84	M	L10/T11	2	N/A/CE	52	M	L11/T12	T; PG3

2.2. Analysis of Specific Gene Expression in Human Primary IVD Cells with Quantitative Polymerase Chain Reaction (qPCR)

Cells were cultured at passage 1–2 (P1–2) in monolayer, and then trypsinised before 2 Mio of cells were lysed directly with TRI. The extraction of total RNA was performed as described in the methods before [28,30]. Prior qPCR RNA integrity and purity were checked on selected samples by the Experion™ Automated Electrophoresis System (Bio-Rad, Reinach, Switzerland). Any possible remaining DNA was digested by using DNase I (AMP-D1, Sigma-Aldrich, Buchs, Switzerland) for 15 min. Reverse transcription was performed by using all-in-one cDNA Synthesis SuperMix (Bimake distributed by LuBio-Science GmbH, Lucerne, Switzerland). Real-time PCR was performed using SYBR Green PCR master mix on a CFX96touch qPCR system (all from Bio-Rad). Ninety genes (Table 2) were tested with PrimePCR™ Pathway plate 96 well (PrimePCR™ TGFβ BMP Signalling Pathway Plus H96, 90 genes, Bio-Rad, US). The qPCR was run using a two-step protocol with an annealing temperature of 60 °C for 30 s and 95 °C for 5 s for 40 cycles (according to the recommendations of the manufacturer). Amplicon specificity was ensured by performing a melting curve analysis after cycling. CFX manager software version 3.1 (Bio-Rad) was used for the quantification analysis of normalised relative gene expression. The number of PCR cycles needed for each sample to reach the threshold level was recorded as the *Cq* value [31]. Values were normalised relative to two reference genes (18S and ACTB). DISH and control cells were compared in scatter plots and RNA transcripts that differed more than four times were indicated in red (up-regulated in DISH patients) or in green (down-regulated in DISH patients).

Table 2. List of genes tested in PrimePCR™ TGFβ BMP Signalling Pathway Plus H96, 90 genes, Bio-Rad, US.

Name	Description	Name	Description	Name	Description
ACTB	Actin beta	*EMP1*	Epithelial Membrane Protein 1	*NOG*	Noggin
ACTC1	Actin Alpha Cardiac Muscle 1	*ENG*	Endoglin	*NOV*	Nephroblastoma overexpressed
ACVR1	Activin A Receptor Type 1	*FAS*	Fas Cell Surface Death Receptor	*PDGFB*	Platelet-Derived Growth Factor Subunit B
ACVR2A	Activin A Receptor Type 2A	*FGF2*	Fibroblast Growth Factor 2	*PLAU*	Plasminogen Activator, Urokinase
ACVRL1	Activin A Receptor-Like Type 1	*FSTL3*	Follistatin Like 3	*PMEPA1*	Prostate Transmembrane Protein, Androgen Induced 1

Table 2. *Cont.*

Name	Description	Name	Description	Name	Description
AMH	Anti-Mullerian Hormone	*GADD45B*	Growth Arrest and DNA Damage Inducible Beta	*RPLP0*	Ribosomal Protein Lateral Stalk Subunit P0
AMHR2	Anti-Mullerian Hormone Receptor Type 2	*GAPDH*	Glyceraldehyde-3-Phosphate Dehydrogenase	*SERPINE1*	Serpine Family E Member 1
ATF4	Activating Transcription Factor 4	*GDF2*	Growth Differentiation Factor 2	*SMAD1*	SMAD Family Member 1
B2M	Beta-2-Microglobulin	*GDF3*	Growth Differentiation Factor 3	*SMAD2*	SMAD Family Member 2
BAMBI	BMP and Activin Membrane Bound Inhibitor	*GDF5*	Growth Differentiation Factor 5	*SMAD3*	SMAD Family Member 3
BGLAP	Bone Gamma-Carboxyglutamate Protein	*GDF6*	Growth Differentiation Factor 6	*SMAD4*	SMAD Family Member 4
BMP1	Bone Morphogenetic Protein 1	*GDF7*	Growth Differentiation Factor 7	*SMAD6*	SMAD Family Member 6
BMP2	Bone Morphogenetic Protein 2	*GSC*	Goosecoid Homeobox	*SMAD7*	SMAD Family Member 7
BMP3	Bone Morphogenetic Protein 3	*GUSB*	Glucuronidase Beta	*SMURF1*	SMAD Specific E3 Ubiquitin Protein Ligase 1
BMP4	Bone Morphogenetic Protein 4	*HERPUD1*	Homocystein Inducible ER Protein with Ubiquitin Like Domain 1	*SOX4*	SRY-Box 4
BMP5	Bone Morphogenetic Protein 5	*HPRT1*	Hypoxanthine Phosphoribosyl-transferase 1	*STAT1*	Signal Transducer and Activator of Transcription 1
BMP6	Bone Morphogenetic Protein 6	*ID1*	Inhibitor of DNA Binding 1, HLH Protein	*STK38L*	Serine/Threonine Kinase 23 Like
BMP7	Bone Morphogenetic Protein 7	*ID2*	Inhibitor of DNA Binding 2	*TBP*	TATA-Box Binding Protein
BMPER	BMP Binding Endothelial Regulator	*IFRD1*	Interferon Related Developmental Regulator 1	*TGFB1*	Transforming Growth Factor Beta 1
BMPR1A	Bone Morphogenetic Protein Receptor, Type IA	*IGF1*	Insulin-Like Growth Factor 1	*TGFB1l1*	Transforming Growth Gactor Beta-1-induced Transcript 1
BMPR1B	Bone Morphogenetic Protein Receptor, Type IB	*IGFBP3*	Insulin-Like Growth Factor Binding Protein 3	*TGFB2*	Transforming Growth Factor Beta 2
BMPR2	Bone Morphogenetic Protein Receptor, Type II	*IL6*	Interleukin 6	*TGFB3*	Transforming Growth Factor Beta 3

Table 2. *Cont.*

Name	Description	Name	Description	Name	Description
CDKN1A	Cyclin-Dependent Kinase Inhibitor 1A	*INHA*	Inhibin Subunit Alpha	*TGFBI*	Transforming Growth Factor Beta Induced
CDKN1B	Cyclin-Dependent Kinase Inhibitor 1B	*JUNB*	JunB Proto-Oncogene, AP-1 Transcription Factor Subunit	*TGFBR1*	Transforming Growth Factor Beta Receptor 1
CDKN2B	Cyclin-Dependent Kinase Inhibitor 2B	*KANK4*	KN Motif and Ankyrin Repeat Domains 4	*TGFBR2*	Transforming Growth Factor Beta Receptor 2
CHRD	Chordin	*KLHL24*	Kelch Like Family Member 24	*TGFBR3*	Transforming Growth Factor Beta Receptor 3
COL1A1	Collagen Type I Alpha 1 Chain	*LEFTY1*	Left-Right Determination Factor 1	*TGIF1*	TGFB Induced Factor Homeobox 1
COL1A2	Collagen Type I Alpha 2 Chain	*LTBP1*	Latent Transforming Growth Factor Beta Binding Protein 1	*THBS1*	Thrombospondin 1
DCN	Decorin	*MYC*	MYC Proto-Oncogene, BHLH Transcription Factor	*TNFSF10*	TNF Superfamily Member 10
EGR2	Early Growth Response 2	*NODAL*	Nodal Growth Differentiation Factor	*UBASH3B*	Ubiquitin Associated and SH3 Domain Containing B

2.3. Statistical Analysis and Visualisation of Heat Maps and Clustering

Relative gene expression was quantified between DISH and "control" patients. Statistical significance was tested using a non-parametric Mann–Whitney U test due to relatively low sample sizes using Prism 7.0d for Mac OS X (GraphPad, La Jolla, CA, USA). Non-parametric distribution was assumed. Consequently, statistical significance was determined using the Mann–Whitney U test. A p-value < 0.05 was considered to be significant. Relative gene expression data were normalised using CFX manager software 3.1. Heat mapping and hierarchical clustering was performed in CFX manager based on the degree of similarity of expression for different samples, and a heat map was generated alongside as implemented in CFX manager software. Pairwise scatter plot analysis was performed between DISH tissues versus "control" tissues and relative gene expression differences > 4.0 were highlighted (Supplementary Figure S1).

2.4. Immunocytochemistry

Cells at P1/P2 of frozen DISH donor 1 and control donor 1 (Table 1) were thawed and seeded at a density of 8000 cells/mL in glass chamber slides with removable 12-well silicone chambers for immunofluorescence staining (growth area per well 1.9 cm^2) (Vitaris, Baar, Switzerland). Cells were then cultured in LG-DMEM supplemented with 10% FBS and 1% P/S until they reached 70% confluency. Cells were then fixed in 4% buffered paraformaldehyde (Sigma-Aldrich) for 15 min and stored in 70% ethanol at 4 °C prior to staining. Fixed cells were permeabilized with 0.2% Triton (Sigma-Aldrich) for 15 min. Again, the cells were washed three times with Tris-buffered saline (TBS) for 5 min. Subsequently, cells were blocked with 10% FCS in 0.025% Tris-buffered saline containing Tween 20 (TBS-T) for 30 min at room temperature. After blocking, cells were incubated overnight at 4 °C using a primary antibody (AB) in blocking solution and under continuous agitation (see Table 3). Secondary ABs were applied in 1:200 dilution in 0.025% TBS-T for 3 h at room temperature (Alexa Fluor® 555 Rabbit Anti-Goat IgG (H+L), Alexa Fluor® 555 Goat anti-mouse SFX-kit, Alexa Fluor® 488 Goat anti-Rabbit SFX Kit, Molecular Probes, Life Technologies, Basel, Switzerland). Cells were viewed under a microscope (Leica DMI4000 (Biosystems inc., Muttenz, Switzerland) and LAS AF Software).

Table 3. Primary antibodies used for immunohistochemistry.

Primary AB	Source	Concentration ICC
Mouse GDF5	Polyclonal Goat IgG (Biotechne, Zug, Switzerland)	5 μg/mL
Human IL6	Monoclonal mouse IgG (Biotechne)	16 μg/mL
Human IGF1	Polyclonal Goat IgG (Biotechne)	5 μg/mL
Human BMP2	Polyclonal Rabbit (Novusbio, Zug, Switzerland)	5.75 μg/mL
Human EGR2	Polyclonal Rabbit (Novusbio)	5 μg/mL
Human GDF6	Rabbit polyclonal–C terminal (Abcam, Cambridge, UK)	1:100

3. Results

3.1. Prime PCR of DISH and Traumatic/Degenerative Discs

PrimePCR™ screening for 90 genes of the BMP pathway was initially run. Most strikingly, these data revealed a significant mean up-regulation of *EGR2* (12.8 ± 7.6-fold, $p = 0.0068$, overall mean ± SEM of all comparisons) and of interleukin 6 (*IL6*) (54.0 ± 46.5-fold, $p = 0.0005$) in DISH-IVDs relative to the "control" discs (Figure 1, Supplementary Figure S1). Furthermore, *IGF1* tended to be up-regulated in DISH-IVD donors (174.1 ± 120.6-fold, $p = 0.1704$), although this was non-significant. Growth and Differentiation Factor 5 (*GDF5*) and *GDF6* both tended to be down-regulated in the DISH-IVDs (i.e., −11.5 ± 10.0, $p = 0.26$ and −3.7 ± 3.1 $p = 0.30$, respectively, Figure 1, Supplementary Figure S1). However, these differences were not significant.

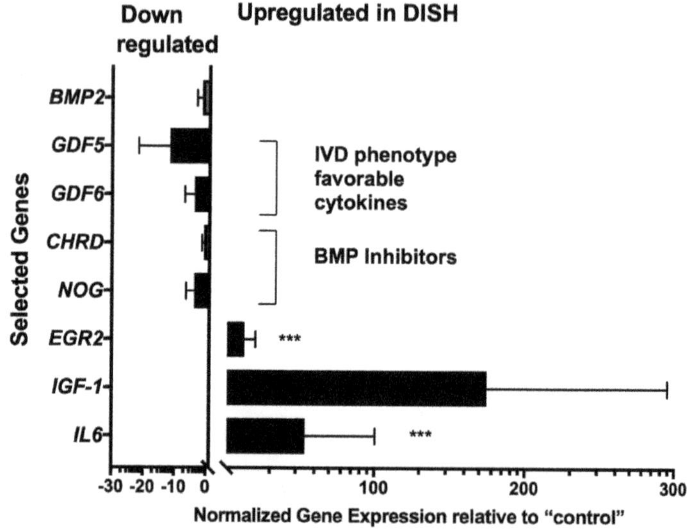

Figure 1. Normalised relative gene expression in DISH-IVDs relative to IVDs of "control" patients. The relative gene expression of IL6 and EGR2 was significantly up-regulated: *** $p < 0.01$. The scatterplots of the pairwise comparisons of the 90 PrimePCR™ gene array is given in Supplementary Figure S1. Means ± SEM.

Clustering the gene expression profiles of the 19 analysed samples of the three control patients and the six DISH patients revealed that not all samples clustered together as expected (Figure 2). However, it seems also evident that the hierarchical phenogram generated two main clusters of phenotypes with a similar TGFβ pathway pattern (Figure 2). Clade 1 consisted of mainly DISH samples, and clade 2 consisted mainly of control samples. However, samples of DISH patients 3 and 5 (i.e., D P3 IVDC and D P5 AFC and EPC) showed a phenotype more similar to the cells of the included trauma patients. On the other hand, the dendrogram confirmed that cells originating from different tissue types of the IVD from the same patient were grouped closely together, which was expected.

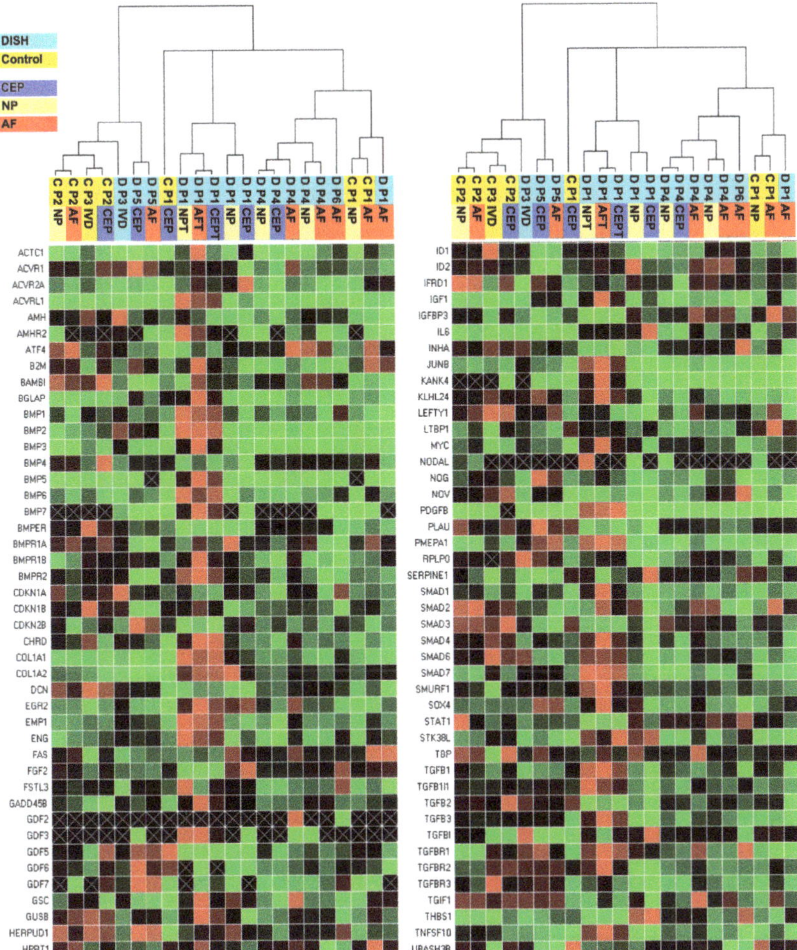

Figure 2. Clustergram displaying hierarchy based on the degree of similarity of expression for different samples and normalised gene expression heat map of relative gene expression at 88 genes of PrimePCR™ of the TGFβ-BMP Signalling Pathway Plus H96. Red indicates up-regulations, green indicates down-regulations and black indicates no regulations. The lighter the shade of colour, the greater the relative normalised expression difference. If no normalised expression value could be calculated, this is shown in black with a white X. Relative gene expression was estimated using ACTB and GAPDH as the two reference genes. Samples encoded in orange are samples categorised as trauma patients − control (C), and samples in light blue are from patients that were diagnosed with DISH (D). Tissue types were colour-coded in dark blue for CEP, in yellow for NP, and in red for AF.

3.2. Immunocytochemistry of IGF1, IL6, EGR2, BMP2, GDF5, and GDF6

Gene expression differences, which showed a more than four times up- or down-regulation in PrimePCR™ between DISH and "control" patients (Figure 1), were further evaluated using immune-staining of isolated cells grown in monolayer. BMP2 showed a stronger signal in the AFC in DISH than in "control" patients, as found in the relative gene expression (Figure 3). Furthermore, the EGR staining was more intense in the AFC and NPC in DISH compared to control patients' staining, as was also found in PrimePCR™. In particular, the AFC showed a strong immunocytochemistry staining for EGR2 and BMP2 compared to the AFC of the control patient. Additionally, IGF1 staining of DISH was more intense compared to control patients, as the NPC and the AFC showed almost no IGF1 protein expression by immunocytochemistry staining. GDF5 and GDF6 were overall more strongly stained in control cells compared to DISH cells.

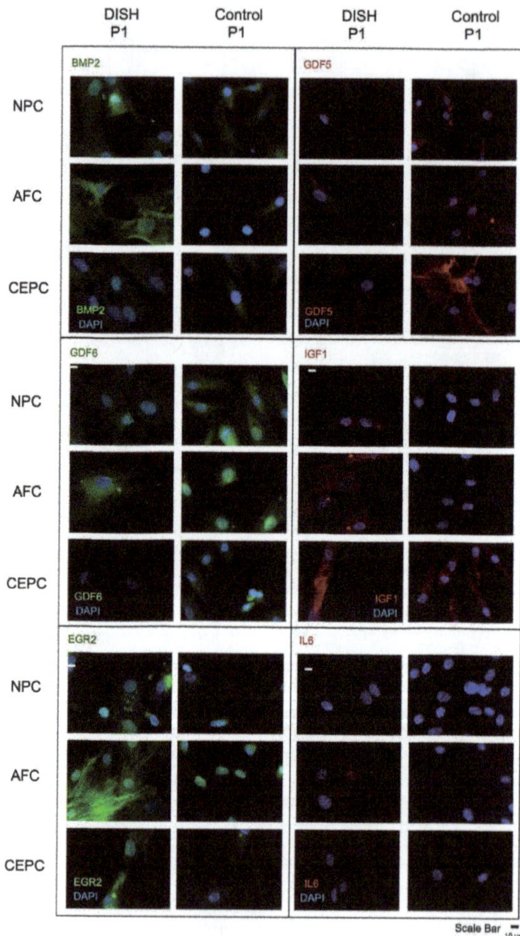

Figure 3. Fluorescence microscopy of immune-histochemistry of nucleus pulposus (NPC), annulus fibrosus (AFC) and cartilaginous endplate cells (CEPC) of DISH versus "control" patient 1 using human GDF5, human IL6, human IGF1, human BMP2, and human EGR2, and human GDF6 antibody and the secondary antibody labeled with the fluorescent dye Alexa 555 (red), and Alexa 488 (green), respectively.

4. Discussion

In this study, we report, for the first time, on the phenotypic changes in IVD cells from DISH patients compared to IVD cells from supposedly "healthier" cells from younger and mid-aged trauma patients (Table 1). We observed the consistent up-regulation of IGF-1, and EGR2, which seems to be a hallmark of the DISH-IVD cells [4,29], and cells from degenerative discs showed a different BMP gene-related transcript and protein expression (Figures 1–3). One of the main limitations of this study was the relatively small sample size. Thus, sample collection from a single hospital was a challenge, and future studies should recruit IVD material from bigger cohorts and multiple populations and etiologies. Additionally, the "controls" used in the study were not truly healthy IVDs, as they were either degenerative or traumatic discs. In this respect, a comparison of fresh cadaveric IVD material of undegenerated nature would be essential for any future investigations. Acquiring more DISH and matching healthy IVDs are required to support our first findings regarding the phenotypic status of the IVD during DISH disease. Moreover, it should be investigated whether DISH affects the overall phenotype of IVD cells and their tissues. To answer this particular question, complete transcriptome analyses using bulk RNA next-generation sequencing would definitively shed new light on the better understanding of DISH pathology and into yet-unidentified modified pathways of this disease.

Since the discovery of DISH, various studies, especially concerning the metabolic conditions of these patients, were conducted. Nevertheless, this disease is still poorly understood. In the current study, we investigated the changes at the transcript and the protein expression level of IVDs from patients diagnosed with DISH. We observed significant up-regulation of EGR2 and IL6 and a trend of higher IGF1 expression at the transcript level. These changes could be confirmed qualitatively at the protein level using immunohistochemistry. For GDF5 and GDF6, we noticed, for both genes, a down-regulation in gene expression in the IVD of those patients, though this was not statistically significant. However, these differences were not apparent at the protein level (Figure 3). This gene expression profile points to an involvement in inflammation in DISH, recently identified in a review by Mader et al. (2021) [29].

Previous research by Levi et al. (1996) [32] has shown that EGR2-/- mice develop skeletal abnormalities, i.e., reducing length and thickness in newly formed bone and calcified trabeculae. These phenomena were explained by the activation of EGR2 in a subpopulation of hypertrophic chondrocytes of the growth plate and differentiating osteoblasts (OBs) [32]. Furthermore, Leclerc et al. (2008) [33] showed up-regulation of EGR2 expression during the late stage of OB differentiation in-vitro. Furthermore, Zaman et al. (2012) [34] found that EGR2 expression was up-regulated in the osteosarcoma-derived UMR106 cell line through IGF1 stimulation. The cause of the observed up-regulation of EGR2 and its role in DISH-IVD cells would need to be further explored.

DISH has been associated with the change in inflammatory mediators [35]. Interleukin 1 (IL1) and IL6 are increased in DISH patients through the activation of nuclear factor kappa-B (NF-κB) ligand, and these cytokines stimulate the proliferation of OBs and bone deposition [29,35,36]. Two other studies [37,38] have also suggested the stimulating effect of IL6 on bone formation and turnover. Kang et al. (1996) [39] further demonstrated an increased concentration of IL6 in herniated lumbar discs, which may indicate an association with degradation of the IVD. We could most interestingly confirm the involvement of IL6 in our data, as it was up-regulated in all analysed DISH-IVDs (Figure 1, and Supplementary Figure S1).

Several studies have shown the possible overlap between DISH, obesity and Diabetes mellitus, with a pathogenic effect of insulin, IGF1 and GH [40–43]. Yakar et al. (2002) [44] demonstrated an association of circulating blood serum IGF1 and bone mass in mice with reduced IGF1 serum levels. These mice showed abnormalities in bone formation including decreased bone mineral density and cortical thickness [44]. However, IGF1 is not only present in the serum but also readily available in the skeletal environment. The skeletal homeostasis is driven by the release of IGF1 during bone modelling and remodelling, as an essential mitogen and chemotactic player in the skeleton [45]. It also could have been shown that IGF1 stimulates IVD cell proliferation and matrix synthesis in-vitro [46].

GDF5, GDF6, and BMP2 are members of the TGFβ superfamily, with GDF5 and GDF6 being involved in the growth and development of cartilage and bone tissues [47]. Moreover, in the IVD, GDF5 seems to play a role in altering homeostasis, as several studies showed its beneficial effect of inducing cell proliferation, proteoglycan stimulation and COL2 expression [48,49]. Furthermore, GDF6 plays a role in spinal column development, especially in IVD development and homeostasis [50,51]. Both GDF5 and GDF6 have recently been shown to promote mesenchymal stromal cells towards a discogenic phenotype [52–56]. Interestingly, both GDF5 and GDF6 were down-regulated in several cell types of DISH-IVDs compared to controls. This might indicate a DISH-related change of the discogenic phenotype in these patients.

With increasing rates of obesity and type 2 Diabetes, the prevalence of DISH is also expected to rise in the future, possibly also affecting younger patients below 50 years of age [29,57]. This development is leading to an urgency to further understand and investigate this disease and its cell signalling.

5. Conclusions

- DISH-IVD, in contrast to IVD obtained from trauma, showed an up-regulation of *EGR2* and *IL-6*, and *IGF-1* tended to be up-regulated.
- Dysregulation of IGF-1 has been proposed for DISH-patients to be a serum-specific marker, and our data from IVD tissue pointed in this direction [4].
- *GDF5* and *GDF6* tended to be down-regulated in DISH-IVD compared to trauma-IVD.
- The observed differences in gene expression and protein differences could also be a result of factors other than DISH such as age and gender bias [58].

Supplementary Materials: The following are available online at https://www.mdpi.com/article/10.3390/10.3390/app11094072/s1: Figure S1: TGFβ pathway transcriptome analysis of three donors with age- and disc level matched controls.

Author Contributions: B.G. and R.D.M. were responsible for the conception and design of the study. R.D.M. performed the experiments, data analysis and prepared the figures. R.D.M. drafted the manuscript; B.G. did extensive editing and prepared all of the figures; P.B.-L. and K.A.C.O. reviewed and edited the manuscript; C.E.A. and L.M.B. provided donor materials and clinical input and contributed to editing. All authors participated in the interpretation of the findings and approved the final version of the manuscript. B.G. and C.E.A. provided funding. All authors have read and agreed to the published version of the manuscript.

Funding: This work was supported from a start-up grant from the Center for Applied Biotechnology and Molecular Medicine (CABMM) to C.E.A. and B.G, and by funds from the Marie Skłodowska Curie International Training Network (ITN) "disc4all" (https://cordis.europa.eu/project/id/955735, accessed on 28 April 2021). Further funds were received from the clinical trials unit of Bern University Hospital and by a Eurospine Task Force Research grant #2019_22.

Institutional Review Board Statement: Not applicable.

Informed Consent Statement: All primary human cells presented in this article were obtained with written consent from the donors. The study was conducted in accordance with the Declaration of Helsinki, and the Ethics Committee approved the protocol of the Canton of Bern (Ref. 2019-00097).

Data Availability Statement: All data presented in this manuscript can be obtained upon request from the corresponding author.

Acknowledgments: We thank Andrea Oberli, Selina Steiner, and Eva Roth for laboratory assistance. The microscopes were provided by the microscope core facility of the University of Bern (www.mic.unibe.ch, accessed 28 April 2021).

Conflicts of Interest: The authors declare no conflict of interest.

Abbreviations

AB	Antibody
AF	Annulus fibrosus
AS	Ankylosing spondylitis
BMP	Bone morphogenetic protein
BMP2	Bone morphogenetic protein 2
BSA	Bovine serum albumin
C	Cervical
CEP	Cartilaginous endplate
D	Degenerated
DISH	Diffuse idiopathic skeletal hyperostosis
EGR2	Early growth response 2
GDF5	Growth and differentiation factor 5
GDF6	Growth and differentiation factor 6
GH	Growth hormone
IGF1	Insulin-like growth factor 1
IL1	Interleukin 1
IL6	Interleukin 6
IVD	Intervertebral disc
L	Lumbar
LG-DMEM	Low-glucose Dulbecco's Modified Eagle Medium
NP	Nucleus pulposus
OB	Osteoblast
TBS-T	Tris-buffered saline containing Tween 20
TGF-β	Transforming growth factor β
P	Passage
PG	Pfirrmann grade
qPCR	Quantitative real-time Polymerase Chain Reaction
T	Trauma

References

1. Forestier, J.; Rotes-Querol, J. Senile Ankylosing Hyperostosis of the Spine. *Ann. Rheum. Dis.* **1950**, *9*, 321–330. [CrossRef] [PubMed]
2. Olivieri, I.; D'Angelo, S.; Palazzi, C.; Padula, A.; Mader, R.; Khan, M.A. Diffuse idiopathic skeletal hyperostosis: Differentiation from ankylosing spondylitis. *Curr. Rheumatol. Rep.* **2009**, *11*, 321–328. [CrossRef] [PubMed]
3. Cammisa, M.; De Serio, A.; Guglielmi, G. Diffuse idiopathic skeletal hyperostosis. *Eur. J. Radiol.* **1998**, *27*, S7–S11. [CrossRef]
4. Mader, R.; Verlaan, J.-J.; Buskila, D. Diffuse idiopathic skeletal hyperostosis: Clinical features and pathogenic mechanisms. *Nat. Rev. Rheumatol.* **2013**, *9*, 741–750. [CrossRef] [PubMed]
5. Resnick, D.; Shaul, S.R.; Robins, J.M. Diffuse Idiopathic Skeletal Hyperostosis (DISH): Forestier's Disease with Extraspinal Manifestations. *Radiology* **1975**, *115*, 513–524. [CrossRef] [PubMed]
6. Pillai, S.; Littlejohn, G. Metabolic Factors in Diffuse Idiopathic Skeletal Hyperostosis—A Review of Clinical Data. *Open Rheumatol. J.* **2014**, *8*, 116–128. [CrossRef] [PubMed]
7. Kiss, C.; Szilágyi, M.; Paksy, A.; Poór, G. Risk factors for diffuse idiopathic skeletal hyperostosis: A case–control study. *Rheumatology* **2002**, *41*, 27–30. [CrossRef]
8. Weinfeld, R.M.; Olson, P.N.; Maki, D.D.; Griffiths, H.J. The prevalence of diffuse idiopathic skeletal hyperostosis (DISH) in two large American Midwest metropolitan hospital populations. *Skelet. Radiol.* **1997**, *26*, 222–225. [CrossRef]
9. Utsinger, P.D. Diffuse idiopathic skeletal hyperostosis. *Clin. Rheum. Dis.* **1985**, *11*, 325–351.
10. Resnick, D.; Guerra, J.; Robinson, C.A.; Vint, V.C. Association of diffuse idiopathic skeletal hyperostosis (DISH) and calcification and ossification of the posterior longitudinal ligament. *Am. J. Roentgenol.* **1978**, *131*, 1049–1053. [CrossRef]
11. Mader, R. Diffuse idiopathic skeletal hyperostosis: A distinct clinical entity. *Isr. Med. Assoc. J.* **2003**, *5*, 506–508.
12. Kuperus, J.S.; Westerveld, L.A.; Rutges, J.A.; Alblas, J.; Van Rijen, M.H.; Bleys, R.L.; Oner, F.C.; Verlaan, J. Histological characteristics of diffuse idiopathic skeletal hyperostosis. *J. Orthop. Res.* **2016**, *35*, 140–146. [CrossRef]
13. El Miedany, Y.M.; Wassif, G.; El Baddini, M. Diffuse idiopathic skeletal hyperostosis (DISH): Is it of vascular aetiology? *Clin. Exp. Rheumatol.* **2000**, *18*, 193–200.
14. Charles, W.D. Growth hormone and insulin-like growth factor-I in symptomatic and asymptomatic patients with diffuse idiopathic skeletal hyperostosis (DISH). *Front. Biosci.* **2002**, *7*, a37–a43. [CrossRef]

15. Denko, C.W.; Boja, B.; Moskowitz, R.W. Growth factors, insulin-like growth factor-1 and growth hormone, in synovial fluid and serum of patients with rheumatic disorders. *Osteoarthr. Cartil.* **1996**, *4*, 245–249. [CrossRef]
16. Matteucci, B.M. Metabolic and endocrine disease and arthritis. *Curr. Opin. Rheumatol.* **1995**, *7*, 356–358. [CrossRef]
17. Daragon, A.; Mejjad, O.; Czernichow, P.; Louvel, J.P.; Vittecoq, O.; Durr, A.; Le Loet, X. Vertebral hyperostosis and diabetes mellitus: A case-control study. *Ann. Rheum. Dis.* **1995**, *54*, 375–378. [CrossRef]
18. Littlejohn, G.O.; Hall, S. Diffuse idiopathic skeletal hyperostosis and new bone formation in male gouty subjects. A radiologic study. *Rheumatol. Int.* **1982**, *2*, 83–86. [CrossRef]
19. Vezyroglou, G.; Mitropoulos, A.; Antoniadis, C. A metabolic syndrome in diffuse idiopathic skeletal hyperostosis. A controlled study. *J. Rheumatol.* **1996**, *23*, 672–676.
20. He, X.; Liu, Y.; Yuan, X.; Lu, L. Enhanced Healing of Rat Calvarial Defects with MSCs Loaded on BMP-2 Releasing Chitosan/Alginate/Hydroxyapatite Scaffolds. *PLoS ONE* **2014**, *9*, e104061. [CrossRef]
21. Brazil, D.P.; Church, R.H.; Surae, S.; Godson, C.; Martin, F. BMP signalling: Agony and antagony in the family. *Trends Cell Biol.* **2015**, *25*, 249–264. [CrossRef]
22. Urist, M.R. Bone: Formation by autoinduction. *Science* **1965**, *150*, 893–899. [CrossRef]
23. Ono, K.; Yonenobu, K.; Miyamoto, S.; Okada, K. Pathology of Ossification of the Posterior Longitudinal Ligament and Ligamentum Flavum. *Clin. Orthop. Relat. Res.* **1999**, *359*, 18–26. [CrossRef]
24. Carragee, E.J.; Hurwitz, E.L.; Weiner, B.K. A critical review of recombinant human bone morphogenetic protein-2 trials in spinal surgery: Emerging safety concerns and lessons learned. *Spine J.* **2011**, *11*, 471–491. [CrossRef]
25. Kok, D.; Peeters, C.M.M.; Mardina, Z.; Oterdoom, D.L.M.; Bulstra, S.K.; Veldhuizen, A.G.; Kuijer, R.; Wapstra, F.H. Is remaining intervertebral disc tissue interfering with bone generation during fusion of two vertebrae? *PLoS ONE* **2019**, *14*, e0215536. [CrossRef]
26. Brown, S.J.; Turner, S.A.; Balain, B.S.; Davidson, N.T.; Roberts, S. Is Osteogenic Differentiation of Human Nucleus Pulposus Cells a Possibility for Biological Spinal Fusion? *Cartilage* **2018**, *11*, 181–191. [CrossRef]
27. Makino, T.; Tsukazaki, H.; Ukon, Y.; Tateiwa, D.; Yoshikawa, H.; Kaito, T. The Biological Enhancement of Spinal Fusion for Spinal Degenerative Disease. *Int. J. Mol. Sci.* **2018**, *19*, 2430. [CrossRef]
28. Tekari, A.; May, R.D.; Frauchiger, D.A.; Chan, S.; Benneker, L.M.; Gantenbein, B. The BMP2 variant L51P restores the osteogenic differentiation of human mesenchymal stromal cells in the presence of intervertebral disc cells. *Eur. Cells Mater.* **2017**, *33*, 197–210. [CrossRef]
29. Mader, R.; Pappone, N.; Baraliakos, X.; Eshed, I.; Sarzi-Puttini, P.; Atzeni, F.; Bieber, A.; Novofastovski, I.; Kiefer, D.; Verlaan, J.-J.; et al. Diffuse Idiopathic Skeletal Hyperostosis (DISH) and a Possible Inflammatory Component. *Curr. Rheumatol. Rep.* **2021**, *23*, 1–6. [CrossRef]
30. Reno, C.; Marchuk, L.; Sciore, P.; Frank, C.; Hart, D. Rapid Isolation of Total RNA from Small Samples of Hypocellular, Dense Connective Tissues. *Biotechniques* **1997**, *22*, 1082–1086. [CrossRef]
31. Ginzinger, D.G. Gene quantification using real-time quantitative PCR: An emerging technology hits the mainstream. *Exp. Hematol.* **2002**, *30*, 503–512. [CrossRef]
32. Levi, G.; Topilko, P.; Schneider-Maunoury, S.; Lasagna, M.; Mantero, S.; Cancedda, R.; Charnay, P. Defective bone formation in Krox-20 mutant mice. *Development* **1996**, *122*, 113–120. [CrossRef] [PubMed]
33. Leclerc, N.; Noh, T.; Cogan, J.; Samarawickrama, D.B.; Smith, E.; Frenkel, B. Opposing effects of glucocorticoids and Wnt signaling on Krox20 and mineral deposition in osteoblast cultures. *J. Cell. Biochem.* **2008**, *103*, 1938–1951. [CrossRef] [PubMed]
34. Zaman, G.; Sunters, A.; Galea, G.L.; Javaheri, B.; Saxon, L.K.; Moustafa, A.; Armstrong, V.J.; Price, J.S.; Lanyon, L.E. Loading-related Regulation of Transcription Factor EGR2/Krox-20 in Bone Cells Is ERK1/2 Protein-mediated and Prostaglandin, Wnt Signaling Pathway-, and Insulin-like Growth Factor-I Axis-dependent. *J. Biol. Chem.* **2012**, *287*, 3946–3962. [CrossRef]
35. Verlaan, J.-J.; Boswijk, P.F.; de Ru, J.A.; Dhert, W.J.; Oner, F.C. Diffuse idiopathic skeletal hyperostosis of the cervical spine: An underestimated cause of dysphagia and airway obstruction. *Spine J.* **2011**, *11*, 1058–1067. [CrossRef]
36. Sarzi-Puttini, P.; Atzeni, F. New developments in our understanding of DISH (diffuse idiopathic skeletal hyperostosis). *Curr. Opin. Rheumatol.* **2004**, *16*, 287–292. [CrossRef]
37. Kobayashi, S.; Momohara, S.; Ikari, K.; Mochizuki, T.; Kawamura, K.; Tsukahara, S.; Nishimoto, K.; Okamoto, H.; Tomatsu, T. A case of Castleman's disease associated with diffuse idiopathic skeletal hyperostosis and ossification of the posterior longitudinal ligament of the spine. *Mod. Rheumatol.* **2007**, *17*, 418–421. [CrossRef]
38. De Benedetti, F.; Rucci, N.; Del Fattore, A.; Peruzzi, B.; Paro, R.; Longo, M.; Vivarelli, M.; Muratori, F.; Berni, S.; Ballanti, P.; et al. Impaired skeletal development in interleukin-6–transgenic mice: A model for the impact of chronic inflammation on the growing skeletal system. *Arthritis Rheum.* **2006**, *54*, 3551–3563. [CrossRef]
39. Kang, J.D.; Georgescu, H.I.; McIntyre-Larkin, L.; Stefanovic-Racic, M.; Donaldson, W.F.; Evans, C.H. Herniated Lumbar Intervertebral Discs Spontaneously Produce Matrix Metalloproteinases, Nitric Oxide, Interleukin-6, and Prostaglandin E2. *Spine* **1996**, *21*, 271–277. [CrossRef]
40. Sahin, G.; Polat, G.; Bagis, S.; Milcan, A.; Erdogan, C. Study of Axial Bone Mineral Density in Postmenopausal Women with Diffuse Idiopathic Skeletal Hyperostosis Related to Type 2 Diabetes Mellitus. *J. Women's Health* **2002**, *11*, 801–804. [CrossRef]
41. Julkunen, H.; Heinonen, O.P.; Pyorala, K. Hyperostosis of the spine in an adult population. Its relation to hyperglycaemia and obesity. *Ann. Rheum. Dis.* **1971**, *30*, 605–612. [CrossRef]

42. Hájková, J.; Streda, A. Hyperostotic spondylosis in women with diabetes mellitus. *Radiol. Diagn.* **1973**, *14*, 71–80.
43. Kiss, C.; O'Neill, T.; Mituszova, M.; Szilágyi, M.; Poór, G. The prevalence of diffuse idiopathic skeletal hyperostosis in a population-based study in Hungary. *Scand. J. Rheumatol.* **2002**, *31*, 226–229. [CrossRef] [PubMed]
44. Yakar, S.; Rosen, C.J.; Beamer, W.G.; Ackert-Bicknell, C.L.; Wu, Y.; Liu, J.-L.; Ooi, G.T.; Setser, J.; Frystyk, J.; Boisclair, Y.R.; et al. Circulating levels of IGF-1 directly regulate bone growth and density. *J. Clin. Investig.* **2002**, *110*, 771–781. [CrossRef] [PubMed]
45. Kawai, M.; Rosen, C.J. The Insulin-Like Growth Factor System in Bone: Basic and Clinical Implications. *Endocrinol. Metab. Clin. North Am.* **2012**, *41*, 323–333. [CrossRef]
46. Osada, R.; Ohshima, H.; Ishihara, H.; Yudoh, K.; Sakai, K.; Matsui, H.; Tsuji, H. Autocrine/paracrine mechanism of insulin-like growth factor-1 secretion, and the effect of insulin-like growth factor-1 on proteoglycan synthesis in bovine intervertebral discs. *J. Orthop. Res.* **1996**, *14*, 690–699. [CrossRef]
47. Storm, E.E.; Huynh, T.V.; Copeland, N.G.; Jenkins, N.A.; Kingsley, D.M.; Lee, S.-J. Limb alterations in brachypodism mice due to mutations in a new member of the TGFβ-superfamily. *Nature* **1994**, *368*, 639–643. [CrossRef]
48. Chujo, T.; An, H.S.; Akeda, K.; Miyamoto, K.; Muehleman, C.; Attawia, M.; Andersson, G.; Masuda, K. Effects of Growth Differentiation Factor-5 on the Intervertebral Disc—In Vitro Bovine Study and In Vivo Rabbit Disc Degeneration Model Study. *Spine* **2006**, *31*, 2909–2917. [CrossRef]
49. Li, X.; Leo, B.M.; Beck, G.; Balian, G.; Anderson, G.D. Collagen and Proteoglycan Abnormalities in the GDF-5-Deficient Mice and Molecular Changes When Treating Disk Cells With Recombinant Growth Factor. *Spine* **2004**, *29*, 2229–2234. [CrossRef]
50. Tassabehji, M.; Fang, Z.M.; Hilton, E.N.; McGaughran, J.; Zhao, Z.; De Bock, C.E.; Howard, E.; Malass, M.; Donnai, D.; Diwan, A.; et al. Mutations in GDF6 are associated with vertebral segmentation defects in Klippel-Feil syndrome. *Hum. Mutat.* **2008**, *29*, 1017–1027. [CrossRef]
51. Hodgkinson, T.; Gilbert, H.T.J.; Pandya, T.; Diwan, A.D.; Hoyland, J.A.; Richardson, S.M. Regenerative Response of Degenerate Human Nucleus Pulposus Cells to GDF6 Stimulation. *Int. J. Mol. Sci.* **2020**, *21*, 7143. [CrossRef]
52. Stoyanov, J.V.; Gantenbein-Ritter, B.; Bertolo, A.; Aebli, N.; Baur, M.; Alini, M.; Grad, S. Role of hypoxia and growth and differentiation factor-5 on differentiation of human mesenchymal stem cells towards intervertebral nucleus pulposus-like cells. *Eur. Cells Mater.* **2011**, *21*, 533–547. [CrossRef]
53. Peroglio, M.; Eglin, D.; Benneker, L.M.; Alini, M.; Grad, S. Thermoreversible hyaluronan-based hydrogel supports in vitro and ex vivo disc-like differentiation of human mesenchymal stem cells. *Spine J.* **2013**, *13*, 1627–1639. [CrossRef]
54. Clarke, L.E.; McConnell, J.C.; Sherratt, M.J.; Derby, B.; Richardson, S.M.; Hoyland, J.A. Growth differentiation factor 6 and transforming growth factor-beta differentially mediate mesenchymal stem cell differentiation, composition, and micromechanical properties of nucleus pulposus constructs. *Arthritis Res. Ther.* **2014**, *16*, R67. [CrossRef]
55. Colombier, P.; Clouet, J.; Boyer, C.; Ruel, M.; Bonin, G.; Lesoeur, J.; Moreau, A.; Fellah, B.-H.; Weiss, P.; Lescaudron, L.; et al. TGF-β1 and GDF5 Act Synergistically to Drive the Differentiation of Human Adipose Stromal Cells towardNucleus Pulposus-like Cells. *STEM CELLS* **2015**, *34*, 653–667. [CrossRef]
56. Miyazaki, S.; Diwan, A.D.; Kato, K.; Cheng, K.; Bae, W.C.; Sun, Y.; Yamada, J.; Muehleman, C.; Lenz, M.E.; Inoue, N.; et al. ISSLS PRIZE IN BASIC SCIENCE 2018: Growth differentiation factor-6 attenuated pro-inflammatory molecular changes in the rabbit anular-puncture model and degenerated disc-induced pain generation in the rat xenograft radiculopathy model. *Eur. Spine J.* **2018**, *27*, 739–751. [CrossRef]
57. Mader, R.; Lavi, I. *Diabetes mellitus* and hypertension as risk factors for early diffuse idiopathic skeletal hyperostosis (DISH). *Osteoarthr. Cartil.* **2009**, *17*, 825–828. [CrossRef]
58. Denko, C.W.; Malemud, C.J. Role of the Growth Hormone/Insulin-like Growth Factor-1 Paracrine Axis in Rheumatic Diseases. *Semin. Arthritis Rheum.* **2005**, *35*, 24–34. [CrossRef]

applied sciences

Article

Diagnostic Limitations and Aspects of the Lumbosacral Transitional Vertebrae (LSTV)

Franz Landauer [1] and Klemens Trieb [1,2,*]

1 Department of Orthopaedic and Trauma Surgery, Paracelsus Medical University Salzburg, 5020 Salzburg, Austria
2 Division for Orthopaedics and Traumatology, Center for Clinical Medicine, Department of Medicine, Faculty of Medicine and Dentistry, Danube Private University, 3500 Krems, Austria
* Correspondence: k.trieb@salk.at

Abstract: The regeneration of an intervertebral disc can only be successful if the cause of the degeneration is known and eliminated. The lumbosacral transitional vertebrae (LSTV) offer itself as a model for IVD (intervertebral disc) regeneration. The aim of this work is to support this statement. In our scoliosis outpatient clinic, 1482 patients were radiologically examined, and ambiguous lumbosacral junction underwent MRI examination. Patients with Castellvi classification type II–IV were included and the results are compared with the current literature in PubMed (12 October 2022). The LSTV are discussed as a possible IVD model. A total of 115 patients were diagnosed with LSTV Castellvi type II–IV. A Castellvi distribution type IIA (n-55), IIB (n-24), IIIA (n-20), IIIB (n-10) and IV (n-6) can be found. In all, 64 patients (55.7%) reported recurrent low-back pain (LBP). Scoliosis (Cobb angle >10°) was also confirmed in 72 patients (58 female and 14 male) and 56 (75.7%) had unilateral pathology. The wide variation in the literature regarding the prevalence of the LSTV (4.6–35.6%) is reasoned by the doubtful diagnosis of Castellvi type I. The LSTV present segments with reduced to absent mobility and at the same time leads to overload of the adjacent segments. This possibility of differentiation is seen as the potential for a spinal model.

Keywords: lumbosacral; transitional; vertebrae; intervertebral disc; scoliosis; Castellvi classification; lumbalgia

Citation: Landauer, F.; Trieb, K. Diagnostic Limitations and Aspects of the Lumbosacral Transitional Vertebrae (LSTV). *Appl. Sci.* **2022**, *12*, 10830. https://doi.org/10.3390/app122110830

Academic Editor: Benjamin Gantenbein

Received: 2 October 2022
Accepted: 24 October 2022
Published: 25 October 2022

1. Introduction

In the introduction, the state of knowledge of the lumbosacral transitional vertebrae (LSTV) is shown on the basis of the current literature. The degeneration of the intervertebral discs is the focus of the work. The literature is reviewed for links between degeneration and regeneration of the intervertebral discs. Attention is paid to the clinical symptoms of lumbar complaints and radiological criteria.

The lumbosacral transitional vertebrae (LSTV) are a congenital malformation of the spine. It is characterized by enlargement of the L5 transverse process(es), with potential pseudoarticulation or fusion with the sacrum. It leads to altered rotational movement of the lower spine, which gives rise to low-back pain (LBP). A diagnosis of Bertolotti syndrome is given to patients experiencing pain caused by the presence of the LSTV. Estimates of the prevalence of the LSTV in the general population vary widely in the literature, ranging from 4.0% to 35.9%. Apazidis et al. have found that type IA is the most common, with a prevalence of 14.7%. However, type I is generally considered to be clinically insignificant [1].

In the Castellvi classification, only type I uses a dimension of >19 mm. Castellvi types II–IV, on the other hand, are defined by a structural change as pseudarthrosis or ossification. This is important because a differentiation between normal structure without altered biomechanics and pathological structure with expected symptoms of discomfort cannot be justified by a measurement of 19 mm. The scattering of the prevalence, which is repeatedly found in the literature, is only one indication of this (Figures 1 and 2).

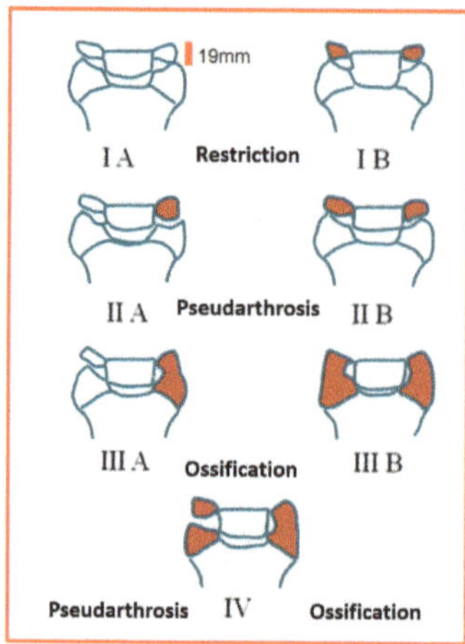

Figure 1. Castellvi classification.

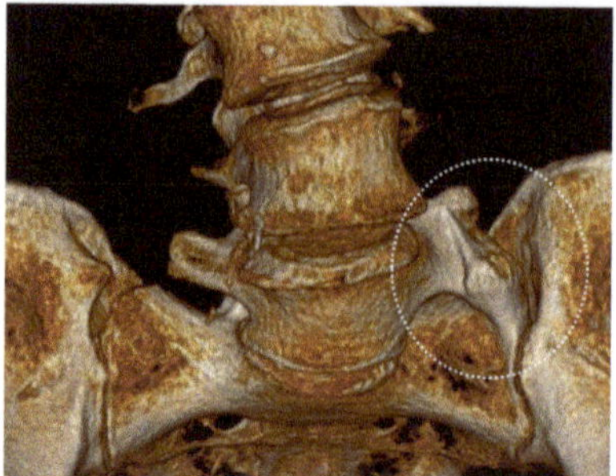

Figure 2. Castellvi IIIA (dotted area shows pathology).

The topic of the inaccuracy of the representation of the L5 transverse process will therefore be discussed in detail at this point. In Figure 1, Castellvi type I is represented as a sketch in such a way that a slight to more severe restriction of the mobility of the affected segment can be assumed (Figure 1). As reflected in the review of the following international literature, this problem has not yet been conclusively addressed. The problem of the imaging quality of the LSTV is presented in a value-neutral manner, even though this topic is repeatedly addressed in the further course. Depending on the pelvic shape and pelvic tilt, the radiological ap image of the lumbosacral junction leads to superimpositions in the case of an enlarged L5 transverse process, which do not always permit measurement and assignment with certainty. The measurement of the L5 transverse process (>19 mm) is thus not guaranteed. As outlined in Figure 3, the imaging result is significantly influenced by the radiological beam path. The imaging technique according to Fergusion with a tube tilt of 30°–40° cranially and caudally suggests improved imaging (Figure 3). However, the dimension of 19 mm remains a variable dependent on many factors (patient body size, tube–object distance, object–film distance, etc.). The pelvic setting in relation to the spine is not only crucial for visualizing pathology, but also an important criterion for possible causes of complaints. Especially the expected disc load in the context of the LSTV is crucial. If now again the question of the cause of complaints in Castellvi type I arises, it becomes clear that a cause differentiation between the LSTV and a pathological pelvic setting is currently hardly possible. No conclusive answer can be found in the literature either. An interaction between the LSTV and a hip complaint can be assumed in individual cases. We therefore always recommend a simultaneous hip examination (Figures 3 and 4).

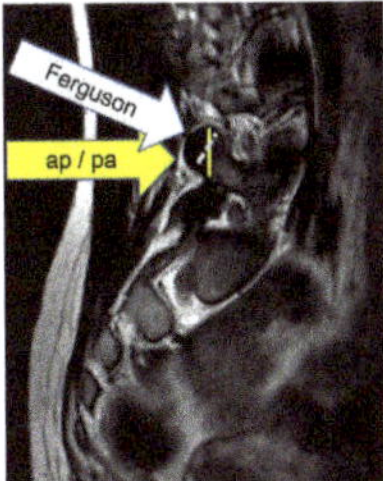

Figure 3. Course of the X-ray beam.

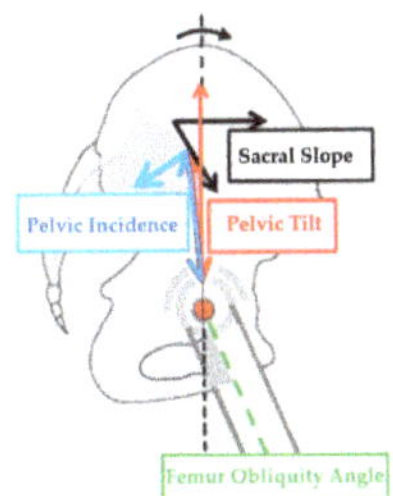

Figure 4. Pelvic setting.

The availability of an MRI examination significantly improves the diagnosis. However, even this examination technique must be specific to the question of the LSTV. Initial referrals of our own to various institutes yielded astonishing results that rarely allowed a diagnosis to be made. This ranged from imaging of only the sagittal and axial slices, which is often sufficient for disc assessment, to imaging that did not portray the L5 transverse process. The combination with a 1.5 Tesla device and a delivered slice thickness of 3 mm also does not allow a more accurate Castellvi type I assessment. These experiences led to arranging for standardized images, as described in the method below, after consultation with one MRI institute. These findings of our own and the still sparse literature at the beginning of the evaluation period led to the decision not to pursue Castellvi type I further for the time being (Figure 5).

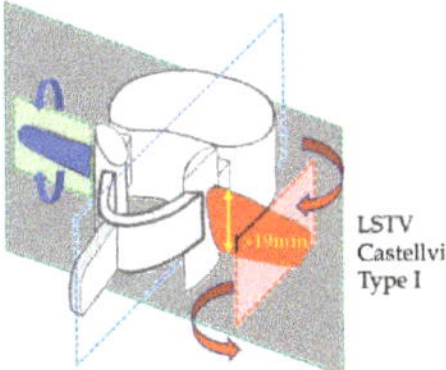

Figure 5. Representation of the L5 transverse process(es) in MRI or CT.

In a special issue on the subject of "Intervertebral Disc Regeneration", the question of the ligamentous situation at the lumbosacral junction must also be considered. The findings available in the literature to date are presented, and the question of the quality of the imaging is part of the problems described above. In particular, the question of mobility restriction of the affected segment and the associated intervertebral disc stress becomes the focus of consideration. This makes it obvious that the Castellvi classification type I is not suitable for differentiating between a normal segment L5/S1 with a pronounced L5 transverse process and a pathological change. This illustration should clarify why the assessment of Castellvi type I was excluded in this study, although these changes are crucial not only for disc degeneration but even more so for the outcome of disc regenerative treatment.

In the work of Min Tang, a prevalence of 15.8% for the LSTV is diagnosed in 5860 radiographs. In all, 44.8% of them are classified as type I [2]. Regarding the question of diagnosis of the LSTV comparing X-ray and MRI, a good agreement was established very early [3].

In the work of the last 10 years, the problem of classification of the LSTV is increasingly worked on. Thus, the imaging quality and reliable assignment of the LSTV, especially of type I, is becoming the focus of discussion.

Farshad et al. advocate measuring segmental differences. Measurement methods in the sagittal slices of MRI images are described for identification of LSTV. However, the variability of the vertebral bodies makes it difficult to accurately define L5 and S1 [4].

Of importance is the total number of vertebral bodies diagnosed with 23 (sacralization) or 25 (lumbarization) with a normal value of 24 vertebral bodies. Thus, a total imaging of the spine is necessary [5].

Quantitative measurements are used to try to find new parameters for the diagnosis and differentiation of the LSTV. Especially LSTV type I with a height of the transverse process of >19 mm forms an uncertainty in the presentation and therefore in the diagnosis. The previously mentioned large differences in the prevalence of the LSTV are partly due to this [6].

In the article by Konin et al., attention to the identification and correct numbering of the LSTV and recognition of imaging findings related to the development of LBP are comprehensively addressed [7].

In previous work, the iliolumbar ligament is recommended for identification of L5/S1 [8]. However, attempts to identify L5/S1 using the iliolumbar ligament have not been successful to date [9].

The iliolumbar ligament in its anatomical structure and as a possible cause of complaints is currently coming into focus. In an MRI study by McGrath et al., unilateral pathology (types IIA and IIIA) shows a thinner ligament than on the unaffected side. Bilateral pathology (types IIB, IIIB, and IV) shows no side difference and no difference from the healthy control group. However, the differentiation of LSTV type I is not answered [10].

For a special issue, "Intervertebral Disc Regeneration II", the cause of LBP and disc degeneration is the focus of the LSTV.

The relationships between IVD degeneration (intervertebral disc degeneration) and the LSTV are differentiated in the literature. In LSTV type I, the highest numbers are reported in some papers; however, no pathological effect on the disc segment is described. In contrast, in LSTV type II–IV, the adjacent cranial disc segment is reported as the segment most exposed to degeneration. The distinction between unilateral and bilateral pathology forms another focus, which is also associated with the question of scoliotic change. A clear prediction cannot be derived so far.

Pathologies of adjacent structures as a cause of complaints are also increasingly subject to clarification. The segmental nervous assignment was examined thereby, too.

The disc degeneration is concentrated on the cranially adjacent segment L4/5. In contrast, the disc of the affected segment of the LSTV type II–IV is mechanically protected by the restriction of movement. However, this general statement still lacks a differentiated assessment dependent on the type of the LSTV. The biomechanical workup of motion restriction and the associated disc loading or unloading are at the beginning of scientific interest [11].

A paper by Ashour et al. states that the incidence of disc herniation is decreased in the segment of the LSTV. In contrast, the incidence of disc herniation or spondylolisthesis in the cranial segment L4-5 is increased in the presence of the LSTV. Therefore, the LSTV are considered a risk factor for degenerative disc changes in the level above the transitional vertebra [12]. In the LSTV type II group, compression of the L5 nerve due to a bony or synovial cause of nearthrosis has also been reported [13]. The question of nerve innervation in the LSTV has also been raised by intraoperative EMG studies, and no pathological distribution was found compared with the normal spine [14]. A significant association between lumbar stenosis and the LSTV is described in a retrospective cross-sectional study [15]. The bony structures show altered trabecular distribution in the LSTV compared to healthy spine in a cadaveric study [16]. In a prospective cadaver study by Golubovsky et al. the segmental mobility of the affected segment and adjacent segments was measured and described as follows:

The results of this study suggest that patients with Bertolotti syndrome have significantly altered spinal biomechanics and may develop pain due to increased loading forces at the LSTV joint during ipsilateral lateral flexion and axial rotation. In addition, increased motion in the upper levels in the presence of the LSTV can lead to degeneration over time [17].

After outlining the pathological changes and movement limitations, the focus turns to the possible causes of LBP:

There are different answers to the question of whether LSTV type I causes symptoms, although diagnostic uncertainty remains as described above. There is also no clear answer to the distinction between unilateral and bilateral changes as the cause of symptoms. However, the more stable the constructs of the LSTV are (type II–IV), the more likely degeneration of the cranially adjacent segment is responsible for the complaints.

Lumbar complaints (LBP) in the presence of the LSTV were originally noted by Mario Bertolotti in 1917 and are therefore referred to as "Bertolotti syndrome" [18].

Quinlan et al. found Bertolotti syndrome in 769 MRI scans of LBP patients in 4.6% of the patients studied. In patients with LBP younger than 30 years, 11.4% showed LSTV [19].

A high frequency of the LSTV was found in young male patients with LBP, and it was mostly subtype I. LSTV type I and associated disc and facet degeneration were notable in this group [20].

A review of the literature on the problem of discomfort in the LSTV (Bertolotti syndrome) shows that comparable changes are found with and without discomfort.

LSTVs pose a diagnostic dilemma for the treating physician, as they may remain undetected on plain radiographs and even on advanced imaging; moreover, even if the malformation is recognized, patients with the LSTV may be asymptomatic or have nonspecific symptoms, such as low-back pain with or without radicular symptoms.

Special attention should be given to younger patients under 30 years of age with chronic lumbar symptoms. Here, diagnostic exclusion or confirmation of the LSTV is required [21].

In lumbalgia, differences in muscle mass distribution are also associated with lordosis of the lumbar spine [22]. The LSTV are associated with a reduction in muscle volume and an increase in muscle degeneration of both the lumbar and trunk muscles compared to healthy individuals [23]. A differential relationship between lumbalgia and the type of the LSTV can be established. However, the segment affected by the LSTV must be distinguished from the adjacent segment as the cause of the symptoms [24]. Unilateral LSTV result in asymmetric biomechanical stresses. The side with the extra L5/S1 joint bears a greater proportion of the load, resulting in lateral tilt with scoliotic curvature. This local loading leads to greater wear and unilateral muscle activity. In addition, asymmetric motion may also affect disc degeneration [25]. The LSTV may also be associated with spina bifida occulta as another pathological change [26,27].

The topic of scoliosis is addressed in 16 papers. In a study in the Turkish Army, the LSTV were diagnosed in 12.9% of 3132 men [28]. Another regional assignment of prevalence to the southern European population is reported by Vinha A. et al. with 24.9% for the group LSTV Castellvi type II, III and IV [29]. In the work of Garg B. et al., a prevalence of LSTV in 198 patients with scoliosis is found in 18.2% [30].

The change in segmental mobility is measured in a paper by Becker L. et al. and summarized as follows: Patients with the LSTV show reduced range of motion in the transitional segment and a significantly increased distribution of motion to the cranially adjacent segment on flexion–extension radiographs. The increased proportion of mobility in the cranial adjacent segment may explain the higher rates of degeneration within the segment [31]. The influence of the LSTV on pelvic alignment with acetabular orientation is also described by the same group [32].

The impact of the LSTV on surgical interventions is also scientifically discussed but is not the focus of this article. An improvement of the patients' quality of life is described by surgical interventions. This concerns both the lysis of LSTV and the fusion of the segment depending on the degenerative changes [33]. However, surgical problems due to the altered anatomy, especially of the ventral vascular plant, must also be considered in surgical planning [34].

The LSTV have a wide range in diagnosis, especially in type I, which is reflected in the widely scattered prevalence data. The biomechanical limitations and the possible causes of the symptoms are not yet sufficiently differentiated. In particular, there is little information about the intervertebral disc in the LSTV. The aim of this paper is to investigate the diagnosis of the LSTV in relation to LBP. The prevalence of the LSTV in the literature and in our own work should be documented with respect to the problem of diagnosis of LSTV, especially of Castellvi type I.

2. Materials and Methods

The present article follows the current PRISMA criteria from 2020. The structure of the contribution is shown in a flow diagram [35] (Table 1).

Table 1. Structure of the contribution.

(1) Introduction:

Presentation of current literature of LSTV:
Prevalence, diagnosis, Castellvi classification, LBP, Bertolotti syndrome, scoliosis, etc.

Flow diagram to Disc: PubMed (12 October 2022):

LSTV $n = 150$
Prevalence $n = 50$
Castellvi classification $n = 51$
LBP $n = 59$
Bertolotti syndrome $n = 19$
Disc $n = 45$
Scoliosis $n = 16$
Surgery $n = 80$

(2) Materials and Methods:

- Current literature focusing on LSTV
- The own LSTV-data (2014–2021)
- Gender, age, Castellvi classification, total number of vertebral bodies, low-back pain, scoliosis, concomitant pathologies, family history.
- Comparison of international literature with own data
- Discussion of segmental limitation according to Castellvi classification.
- LSTV degeneration mechanism as a model for the mechanical limits of IVD regeneration.

(3) Results:
Results of own LSTV-data (2014–2021) (gender, age, Castellvi classification, total number of vertebral bodies, low-back pain, scoliosis, concomitant pathologies, family history).
Castellvi classification
total number of vertebral bodies
low-back pain (Bertolotti syndrome)
scoliosis
gender
age
concomitant pathologies
family history

(4) Discussion of the literature and own data of LSTV:

- Diagnostic limits of LSTV Castellvi type I
- Functional segmental limitation according to Castellvi classification
- LSTV degeneration mechanism as a model for the mechanical limits of IVD regeneration

(5) Summary:
LSTV as an experimental model for intervertebral disc degeneration/regeneration

Diagnosis of LSTV: In our scoliosis outpatient clinic 1482 patients were examined by one consultant orthopaedic surgeon (2014–2021) with more than 25 years of experience in that field.

Each radiological, structurally ambiguous assignment of the lumbosacral junction was submitted to further MRI examination with a 3.5 Tesla device (sagittal, coronal, axial and adjustment to the segment L5/S1 with a slice thickness of 3 mm, Siemens Medical Group).

Patients with Castellvi I classification (height of L5 transverse process >19 mm) are not included in the study. This decision was made in the initial phase of the study, because in our own unpublished studies the imaging inaccuracy allows an exact measurement of the L5 transverse process only in the Ferguson image. However, this additional X-ray is not permitted for reasons of radiation hygiene. On the other hand, 19 mm should be evaluated differently for a very tall person than for a short person.

Only patients with Castellvi classification type II–IV are included in the study.

3. Results

In 1482 spine patients examined, 115 patients were diagnosed with LSTV Castellvi II–IV (male n-32, female n-83).

In the Castellvi IIA (n-55) and IIB (n-24) groups, spondylolysis of the adjacent segment was seen in two patients and lumbarization in one patient, respectively. In patients with Castellvi IIIA (n-20), spondylolysis of the adjacent segment was found in three patients and there was lumbarization in one patient. In classification IIIB (n-10), spondylolisthesis (Meyerding 2) was seen. Six patients showed Castellvi type IV.

A total of 64 (55.7%) of all patients reported recurrent LBP. Of these, 23 patients showed orthograde lumbosacral transition and 41 patients showed scoliosis.

In unilateral pathology (IIA, IIIA, n-75), 39 complained of symptoms, and in bilateral pathology (IIB, IIIB, IV, n-40), 26 patients did. Only five patients belong to group IIIA and thus have neither unilateral nor pseudarthrosis as cause of pain.

In lumbalgia, a clear age-dependent differentiation is evident. In children and adolescents <16 years, 26.7% complained of chronic complaints. In the age group 16–30 years, it was 35.5%, and in >30 years it was 88.6% who reported chronic complaints.

Five patients were therefore already in psychological treatment. In the anamnesis, up to 30 visits to the doctor for lumbalgia can be recorded up to the time of the diagnosis of LSTV. Three patients with recurrent lumbalgia have already undergone spinal surgery without the LSTV being known to the patients.

In 72 patients, LSTV Castellvi II–IV and scoliosis was confirmed (58 female and 14 male). The curvature extent of scolioses showed a mean Cobb angel of 24.3° with a range from 11°–55° (n-42) in children and in adults a mean of 32.4° with a range from 12°–66° (n-30) in the adults.

Of the 72 patients with scoliosis, 56 showed unilateral pathology (IIA and IIIA) and 16 bilateral pathologies (IIB, IIIB, IV, Table 2).

A positive family history existed in 20 patients from the scoliosis group (n = 72). In one family, the mother and daughter were affected type IIA, and in another case it was twins with IIB with scoliotic component. Other malformations were raised: longitudinal malformation of the right upper extremity, spina bifida, other vertebral body malformations (n-3), a sacral cyst, syringomyelia (n-2) and a fork rib.

When comparing our own data with the results in the literature, it should be noted that in our own data only LSTV Castellvi type II, III and IV are collected. The data from the literature refer to the total sum of LSTV Castellvi type I–IV. If Castellvi type I is shown in the data, the value is given in parentheses (Table 3).

Table 2. Data of the own investigation.

	Spine patients (all)				n-1482
	LSTV (all)	♀n-83	♂n-32		n-115
Castellvi	IIA	n-55		IIB	n-24
	IIIA	n-20		IIIB	n-10
				IV	n-6
	Lumbalgia (all)	♀n-46	♂n-18		n-64
Castellvi	IIA	n-30		IIB	n-13
	IIIA	n-11		IIIB	n-5
				IV	n-5
	Scoliosis (all)	♀n-58	♂n-14		n-72
Castellvi	IIA	n-42		IIB	n-11
	IIIA	n-14		IIIB	n-3
				IV	n-2

Table 3. Own data in comparison to the literature.

	Own Data Type II–IV	**Literature Type I–IV**	
LSTV total	7.8%	20.9% (+14.7% Type I)	[1]
		15.8%	[2]
		10.7% (+21.3% Type I)	[20]
		21.1%	[24]
		12.9%	[28]
		9.3% (+19.7% Type I)	[36]
		31%	[37]
LBP total	4.3%	4.6%	[19]
Bertolotti syndrome	55.7%	40.0%	[13]
		30%	[11]
		24.9%	[29]
Scoliosis and LSTV	62.6%	18.2%	[30]

4. Discussion

Castellvi type I:

In the initial phase of the examination, the Castellvi type I problem with its enlargement of the L5 transverse process of >19 mm becomes obvious. The standard radiograph of the lumbar spine, as well as the spinal X-ray for scoliosis clarification, does not result in a reliable orthograde representation of the L5 transverse process. Lordosis and pelvic tilt, as well as scoliosis, prevent standardized imaging. The magnification factor that inevitably occurs in a radiograph as a function of object–film distance and tube distance make the height indication of the L5 transverse process an unsuitable measurement for Castellvi I classification.

Only the Ferguson image (30° angle AP radiograph) serves as a standard method for detecting LSTV. Additional imaging is prohibited for radiation hygiene reasons. However, the problem of the magnification factor remains.

The alternative is an MRI examination, which provides additional information in all three planes in space. The standard images for disc assessment (coronal and sagittal) are not sufficient for the assessment of LSTV. The frequently used slice thickness of 3 mm and the quality of the device (<3.5 Tesla) form another uncertainty factor for diagnostics. If in the literature there are reported (in many cases not specified) investigations with a CT and MRI slice thickness of 0.6 mm–3 mm, this should not be considered an insignificant representation and therefore also not an inaccurate measurement [3,4,6,10].

However, the height of the L5 transverse process must also depend on the overall size of the patient, so that a measurement >19 mm is only of limited use for classification.

In conclusion, the shape of the L5 transverse process could be more indicative of pathology than the measurement alone.

In our own work, because of the intractability of the exact classification of Castellvi type I, we did not work on this type further. This does not mean that Castellvi type I cannot be responsible for complaints [8–10]. It should be emphasized that the scoliosis outpatient clinic, with its patients and the frequent oblique position of L5 compared to S1, presents an additional problem of presentation [1,2].

This raises the following questions:

Is pathology of the ligamentous apparatus (iliolumbar ligament) responsible for the shape of the L5 transverse process? If so, the shape of the L5 transvers process would be the expected identifying feature.

A standardized clarification of the ligamentous apparatus and appropriate classification must be required.

If it is a pathology with an effect on the intervertebral disc, then any interventional procedure on the intervertebral disc must also take into account a possible pathology due to LSTV. This assessment, at first sight very critical, becomes crucial when a new treatment modality is aimed at repairing the disc [37–39].

If the incidence of the LSTV indeed reaches values as reported in some literature between 4% and 35.9% and the LSTV leads to treatment failure, this would be crucial for success and failure for this new treatment modality [21].

LSTV type II:

In unilateral or bilateral pseudarthrosis, limitation of motion of the L5/S1 segment is assured. Therefore, the pseudarthrosis and the intervertebral disc must be differentiated as the cause of the complaint. It is to be expected that the pseudarthrosis cannot follow the age-related process of height reduction in the intervertebral disc without complaints. Corresponding complaints are thus to be expected. For this differentiation, further MRI criteria have to be defined.

We would like to point out another aspect to this.

As patients age, degenerative de novo curvatures or increases in curvature occur. This can lead to a contact zone of the L5 transverse process with the os sacrum (os ileum). If there is no previous X-ray from youth, the assignment as Castellvi II is also fraught with uncertainty. For the question of the IVD, which is the focus of this article, this may have no clinical significance. The literature provides the following evidence in this regard:

Type II LSTV had significant effects in promoting transitional and adjacent disc degeneration.

Thus, Castellvi type III/IV LSTV predisposed the adjacent spinal components to degeneration and protected the transitional discs [31,40].

LSTV type III:

In its unilateral form, stress loading is inevitable. In its bilateral form, stability of the L5/S1 segment is assumed. In both cases, L4/L5 overload is expected due to the lack of L5/S1 mobility [17,31].

LSTV type IV:

This is a bilateral form with a bony bridging and pseudarthrosis. In this case, an increased load on the adjacent segment L4/L5 may be assumed. Moreover, the pseudarthrosis may be responsible for discomfort since there is no symmetrical situation.

Disc degeneration:

In a paper by Ahmed Ashour et al., the incidence for disc prolapse is found to be decreased in the segment of LSTV. In contrast, the incidence for disc herniation or spondylolisthesis in the cranial segment L4-5 is increased in the presence of LSTV [12]. Hence, the LSTV are considered a risk factor for disc degenerative changes at the level above the transitional vertebra level. The IVD only makes sense (can be successful) if it does not affect the adjacent segment of the LSTV. Thus, a differential diagnosis of the LSTV is important before any IVD intervention.

Low-Back Pain (Bertolotti syndrome):

Explanation of own data:

A total of 64 patients of the total group (55.7%) reported recurrent LBP (orthograde lumbar spine $n = 23$, scoliosis $n = 41$).

The low number of LBP can be explained by the fact that the age limit of the patients is very low (scoliosis outpatient clinic). Additionally, the differentiation according to age shows a rapid increase in complaints in the patient distribution, and the dominance of scoliosis with 64.1% is striking.

Basis for an intervertebral disc model with different calculable limitation of motion:

A prospective cadaveric study measured the segmental range of motion of the affected segment and adjacent segments.

A biomechanical analysis of the tissue, joints and associated motion restriction (range of motion, resulting torques and the LSTV joint forces) and histological assessment of the segments was performed. The aim of this work is surgical therapy.

This study's results indicate that patients with Bertolotti syndrome have significantly altered spinal biomechanics and may develop pain due to increased loading forces at the LSTV joint with ipsilateral lateral bending and axial rotation. In addition, increased motion at superior levels when the LSTV are present may lead to degeneration over time.

However, the model can be adapted to address disc treatment issues.

The LSTV are a model given in nature for a differential study of motion restriction on the affected and adjacent disc segments [17,31].

The preceding illustration suggests the LSTV as a potential model for IVD research. The LSTV provide segments with reduced to absent mobility and at the same time an overloaded adjacent segment.

MRI imaging capabilities should differentiate between complaint origin in the LSTV and disc degeneration.

The model representation should also be suitable for further FEM calculation [36–40].

Disc model:

The work shown here offers the LSTV as a disc model for further investigation. As a model for the degeneration of the lumbosacral junction and thus clarification of the cause of the complaint, treatment options become delineable. It can thus lead to a further gain in knowledge for regenerative disc treatment. We are well aware of the limitations of the model, especially in the area of Castellvi type I. Knowing the limitations of imaging will be an advantage for every practitioner and will lead to the avoidance of frustrated therapy attempts in cases of mechanically persisting pathology.

Limitation of own data:

The data were collected in the spine outpatient clinic with a focus on scoliosis. This explains the high number of scolioses. This is clearly shown in Table 3 in comparison with the literature.

The lack of inclusion of LSTV Castellvi I cases prevents direct comparison with the prevalence data in the literature. A wide dispersion is shown, which cannot be narrowed down by our own work. This will be discussed in detail in the discussion.

Through cooperation with an MRI institute, an attempt was made to provide standardized imaging as compensation and thus to ensure a reliable assignment of the patients LSTV Castellvi II, III and IV.

5. Conclusions

This paper clarifies the diagnosis of the LSTV as a possible cause of LBP. From the prevalence of LSTV, in the literature and in our own work, it becomes clear that this is a clinically non-negligible number of patients. In the wide dispersion of prevalence, the problem of diagnosis of LSTV Castellvi type I becomes apparent.

Despite these aforementioned diagnostic problems, the LSTV offer the full spectrum of possible segmental movement limitations. This ranges from a slight limitation of the range of motion, without clinical relevance, to complete bony fixation. Thus, the cranially adjacent disc segment is also included in a possible functional model of the IVD.

This in vivo IVD model is thus suitable to provide further insights into the degeneration mechanism. Knowing the cause of degeneration is the basic prerequisite for successful regeneration treatment using the IVD. For the therapy decision of an IVD regeneration or a necessary alternative treatment procedure, the presented model can provide information in the future.

Author Contributions: Data curation, F.L.; Investigation, F.L.; Methodology, K.T.; Writing—original draft, F.L.; Writing—review & editing, K.T. All authors have read and agreed to the published version of the manuscript.

Funding: This research received no external funding.

Institutional Review Board Statement: Ethical review and approval were waived for this study because only radiographs were analyzed.

Informed Consent Statement: Not applicable.

Data Availability Statement: Not applicable.

Conflicts of Interest: The authors declare no conflict of interest.

Abbreviations

LSTV	Lumbo Sacral Transitional Vertebra
IVD	InterVertebral Disc
MRI	Magnetic resonance imaging
FEM	Finite Element Model
IDR	Intervertebral Disc Regeneration
LBP	Low-Back Pain

References

1. Apazidis, A.; Ricart, P.A.; Diefenbach, C.M.; Spivak, J.M. The prevalence of transitional vertebrae in the lumbar spine. *Spine J.* **2011**, *11*, 858–862. [CrossRef] [PubMed]
2. Tang, M.; Yang, X.F.; Yang, S.W.; Han, P.; Ma, Y.M.; Yu, H.; Zhu, B. Lumbosacral transitional vertebra in a population-based study of 5860 individuals: Prevalence and relationship to low back pain. *Eur. J. Radiol.* **2014**, *83*, 1679–1682. [CrossRef] [PubMed]
3. O'Driscoll, C.M.; Irwin, A.; Saifuddin, A. Variations in morphology of the lumbosacral junction on sagittal MRI: Correlation with plain radiography. *Skelet. Radiol.* **1996**, *25*, 225–230. [CrossRef]
4. Farshad, M.; Aichmair, A.; Hughes, A.P.; Herzog, R.J.; Farshad-Amacker, N.A. A reliable measurement for identifying a lumbosacral transitional vertebra with a solid bony bridge on a single-slice midsagittal MRI or plain lateral radiograph. *Bone Jt. J.* **2013**, *95*, 1533–1537. [CrossRef]
5. Okamoto, M.; Hasegawa, K.; Hatsushikano, S.; Kobayashi, K.; Sakamoto, M.; Ohashi, M.; Watanabe, K. Influence of lumbosacral transitional vertebrae on spinopelvic parameters using biplanar slot scanning full body stereoradiography-analysis of 291 healthy volunteers. *J. Orthop. Sci.* **2022**, *27*, 751–759. [CrossRef] [PubMed]
6. Zhou, S.; Du, L.; Liu, X.; Wang, Q.; Zhao, J.; Lv, Y.; Yang, H. Quantitative measurements at the lumbosacral junction are more reliable parameters for identifying and numbering lumbosacral transitional vertebrae. *Eur. Radiol.* **2022**, *32*, 5650–5658. [CrossRef]
7. Konin, G.P.; Walz, D.M. Lumbosacral transitional vertebrae: Classification, imaging findings, and clinical relevance. *Am. J. Neuroradiol.* **2010**, *31*, 1778–1786. [CrossRef]
8. Hughes, R.J., Saifuddin, A. Numbering of lumbosacral transitional vertebrae on MRI: Role of the iliolumbar ligaments. *Am. J. Roentgenol.* **2006**, *187*, W59–W65. [CrossRef]

9. Farshad-Amacker, N.A.; Lurie, B.; Herzog, R.J.; Farshad, M. Is the iliolumbar ligament a reliable identifier of the L5 vertebra in lumbosacral transitional anomalies? *Eur. Radiol.* **2014**, *24*, 2623–2630. [CrossRef]
10. McGrath, K.A.; Lee, J.; Thompson, N.R.; Kanasz, J.; Steinmetz, M.P. Identifying the relationship between lumbar sacralization and adjacent ligamentous anatomy in patients with Bertolotti syndrome and healthy controls. *J. Neurosurg. Spine* **2022**, *37*, 200–207. [CrossRef]
11. Zhang, B.; Wang, L.; Wang, H.; Guo, Q.; Lu, X.; Chen, D. Lumbosacral Transitional Vertebra: Possible Role in the Pathogenesis of Adolescent Lumbar Disc Herniation. *World Neurosurg.* **2017**, *107*, 983–989. [CrossRef] [PubMed]
12. Ashour, A.; Hassan, A.; Aly, M.; Nafady, M.A. Prevalence of Bertolotti's Syndrome in Lumbosacral Surgery Procedures. *Cureus* **2022**, *14*, e26341. [CrossRef] [PubMed]
13. Kanematsu, R.; Hanakita, J.; Takahashi, T.; Minami, M.; Tomita, Y.; Honda, F. Extraforaminal entrapment of the fifth lumbar spinal nerve by nearthrosis in patients with lumbosacral transitional vertebrae. *Eur. Spine J.* **2020**, *29*, 2215–2221. [CrossRef] [PubMed]
14. Hinterdorfer, P.; Parsaei, B.; Stieglbauer, K.; Sonnberger, M.; Fischer, J.; Wurm, G. Segmental innervation in lumbosacral transitional vertebrae (LSTV): A comparative clinical and intraoperative EMG study. *J. Neurol. Neurosurg. Psychiatry* **2010**, *81*, 734–741. [CrossRef]
15. Abbas, J.; Peled, N.; Hershkovitz, I.; Hamoud, K. Is Lumbosacral Transitional Vertebra Associated with Degenerative Lumbar Spinal Stenosis? *Biomed. Res. Int.* **2019**, *10*, 2019. [CrossRef]
16. Mahato, N.K. Trabecular bone structure in lumbosacral transitional vertebrae: Distribution and densities across sagittal vertebral body segments. *Spine J.* **2013**, *13*, 932–937. [CrossRef]
17. Golubovsky, J.L.; Colbrunn, R.W.; Klatte, R.S.; Nagle, T.F.; Briskin, I.N.; Chakravarthy, V.B.; Gillespie, C.M.; Reith, J.D.; Jasty, N.; Benzel, E.C.; et al. Development of a novel in vitro cadaveric model for analysis of biomechanics and surgical treatment of Bertolotti syndrome. *Spine J.* **2020**, *20*, 638–656. [CrossRef]
18. Bertolotti, M. Contributo alla conoscenza dei vizi di differenziazione regionale del rachide con speciale riguardo all assimilazione sacrale della v. lombare. *Radiol. Med.* **1917**, *4*, 113–144.
19. Quinlan, J.F.; Duke, D.; Eustace, S. Bertolotti's syndrome: A cause of back pain in young people. *J. Bone Jt. Surg. Br.* **2006**, *88*, 1183–1186. [CrossRef]
20. Apaydin, M.; Uluc, M.E.; Sezgin, G. Lumbosacral transitional vertebra in the young men population with low back pain: Anatomical considerations and degenerations (transitional vertebra types in the young men population with low back pain). *Radiol. Med.* **2019**, *124*, 375–381. [CrossRef]
21. McGrath, K.; Schmidt, E.; Rabah, N.; Abubakr, M.; Steinmetz, M. Clinical assessment and management of Bertolotti Syndrome: A review of the literature. *Spine J.* **2021**, *21*, 1286–1296. [CrossRef] [PubMed]
22. Ulger, F.E.B.; Illeez, O.G. The Effect of Lumbosacral Transitional Vertebrae (LSTV) on Paraspinal Muscle Volume in Patients with Low Back Pain. *Acad. Radiol.* **2020**, *27*, 944–950. [CrossRef] [PubMed]
23. Becker, L.; Ziegeler, K.; Diekhoff, T.; Palmowski, Y.; Pumberger, M.; Schömig, F. Musculature adaption in patients with lumbosacral transitional vertebrae: A matched-pair analysis of 46 patients. *Skelet. Radiol.* **2021**, *50*, 1697–1704. [CrossRef] [PubMed]
24. Hanhivaara, J.; Määttä, J.H.; Karppinen, J.; Niinimäki, J.; Nevalainen, M.T. The Association of Lumbosacral Transitional Vertebrae with Low Back Pain and Lumbar Degenerative Findings in MRI: A Large Cohort Study. *Spine* **2022**, *47*, 153–162. [CrossRef] [PubMed]
25. Mahato, N.K. Lumbosacral transitional vertebrae: Variations in low back structure, biomechanics, and stress patterns. *J. Chiropr. Med.* **2012**, *11*, 134–135. [CrossRef]
26. Morimoto, M.; Sugiura, K.; Higashino, K.; Manabe, H.; Tezuka, F.; Wada, K.; Yamashita, K.; Takao, S.; Sairyo, K. Association of spinal anomalies with spondylolysis and spina bifida occulta. *Eur. Spine J.* **2022**, *31*, 858–864. [CrossRef]
27. Sharma, A.; Kumar, A.; Kapila, A. Co-existence of spina bifida occulta and lumbosacral transitional vertebra in patients presenting with lower back pain. *Reumatologia* **2022**, *60*, 70–75. [CrossRef]
28. Fidan, F.; Çay, N.; Asiltürk, M.; Veizi, E. The incidence of congenital lumbosacral malformations in young male Turkish military school candidates population. *J. Orthop. Sci.* **2021**, *11*, S0949–S2658. [CrossRef]
29. Vinha, A.; Bártolo, J.; Lemos, C.; Cordeiro, F.; Rodrigues-Pinto, R. Lumbosacral transitional vertebrae: Prevalence in a southern European population and its association with low back pain. *Eur. Spine J.* **2022**, 1–7. [CrossRef]
30. Garg, B.; Mehta, N.; Goyal, A.; Rangaswamy, N.; Upadhayay, A. Variations in the Number of Thoracic and Lumbar Vertebrae in Patients With Adolescent Idiopathic Scoliosis: A Retrospective, Observational Study. *Int. J. Spine Surg.* **2021**, *15*, 359–367. [CrossRef]
31. Becker, L.; Schönnagel, L.; Mihalache, T.V.; Haffer, H.; Schömig, F.; Schmidt, H.; Pumberger, M. Lumbosacral transitional vertebrae alter the distribution of lumbar mobility-Preliminary results of a radiographic evaluation. *PLoS ONE* **2022**, *17*, e0274581. [CrossRef] [PubMed]
32. Becker, L.; Taheri, N.; Haffer, H.; Muellner, M.; Hipfl, C.; Ziegeler, K.; Diekhoff, T.; Pumberger, M. Lumbosacral Transitional Vertebrae Influence on Acetabular Orientation and Pelvic Tilt. *J. Clin. Med.* **2022**, *11*, 5153. [CrossRef]
33. McGrath, K.A.; Thompson, N.R.; Fisher, E.; Kanasz, J.; Golubovsky, J.L.; Steinmetz, M.P. Quality-of-life and postoperative satisfaction following pseudoarthrectomy in patients with Bertolotti syndrome. *Spine J.* **2022**, *22*, 1292–1300. [CrossRef] [PubMed]

34. Becker, L.; Amini, D.A.; Ziegeler, K.; Muellner, M.; Diekhoff, T.; Hughes, A.P.; Pumberger, M. Approach-related anatomical differences in patients with lumbo-sacral transitional vertebrae undergoing lumbar fusion surgery at level L4/5. *Arch. Orthop. Trauma Surg.* **2022**, 1–7. [CrossRef] [PubMed]
35. Page, M.J.; McKenzie, J.E.; Bossuyt, P.M.; Boutron, I.; Hoffmann, T.C.; Mulrow, C.M.; Shamseer, L.; Tetzlaff, J.M.; Akl, E.A.; Brennan, S.E.; et al. The PRISMA 2020 statement: An updated guideline for reporting systematic reviews. *BMJ* **2021**, *372*, n71. [CrossRef]
36. Hanhivaara, J.; Määttä, J.H.; Niinimäki, J.; Nevalainen, M.T. Lumbosacral transitional vertebrae are associated with lumbar degeneration: Retrospective evaluation of 3855 consecutive abdominal CT scans. *Eur. Radiol.* **2020**, *30*, 3409–3416. [CrossRef]
37. Griffith, J.F.; Xiao, F.; Hilkens, A.; Han, I.; Griffith, Y.; Chi, J.; Leung, S. Increased vertebral body area, disc and facet joint degeneration throughout the lumbar spine in patients with lumbosacral transitional vertebrae. *Eur. Radiol.* **2022**, *32*, 6238–6246. [CrossRef]
38. Guerrero, J.; Häckel, S.; Croft, A.S.; Hoppe, S.; Albers, C.E.; Gantenbein, B. The nucleus pulposus microenvironment in the intervertebral disc: The fountain of youth? *Eur. Cell. Mater.* **2021**, *41*, 707–738. [CrossRef]
39. Bermudez-Lekerika, P.; Crump, K.B.; Tseranidou, S.; Nüesch, A.; Kanelis, E.; Alminnawi, A.; Baumgartner, L.; Muñoz-Moya, E.; Compte, R.; Gualdi, F.; et al. Immuno-Modulatory Effects of Intervertebral Disc Cells. *Front. Cell Dev. Biol.* **2022**, *10*, 924692. [CrossRef]
40. Cheng, L.; Jiang, C.; Huang, J.; Jin, J.; Guan, M.; Wang, Y. Lumbosacral Transitional Vertebra Contributed to Lumbar Spine Degeneration: An MR Study of Clinical Patients. *J. Clin. Med.* **2022**, *11*, 2339. [CrossRef]

 *applied sciences*

Article

Using Natural Language Processing to Identify Low Back Pain in Imaging Reports

Yeji Kim [1,†], Chanyoung Song [2,†], Gyuseon Song [2], Sol Bi Kim [3], Hyun-Wook Han [2,*] and Inbo Han [3,*]

1 Research Competency Milestones Program of School of Medicine, CHA University School of Medicine, Bundang-gu, Seongnam-si 13488, Republic of Korea
2 Department of Biomedical Informatics, CHA University School of Medicine, Bundang-gu, Seongnam-si 13488, Republic of Korea
3 Department of Neurosurgery, CHA University School of Medicine, CHA Bungdang Medical Center, Seongnam-si 13497, Republic of Korea
* Correspondence: hwhan_dj@chamc.co.kr (H.-W.H.); hanib@cha.ac.kr (I.H.); Tel.: +82-31-780-1924 (I.H.)
† These authors contributed equally to this work.

Abstract: A natural language processing (NLP) pipeline was developed to identify lumbar spine imaging findings associated with low back pain (LBP) in X-radiation (X-ray), computed tomography (CT), and magnetic resonance imaging (MRI) reports. A total of 18,640 report datasets were randomly sampled (stratified by imaging modality) to obtain a balanced sample of 300 X-ray, 300 CT, and 300 MRI reports. A total of 23 radiologic findings potentially related to LBP were defined, and their presence was extracted from radiologic reports. In developing NLP pipelines, section and sentence segmentation from the radiology reports was performed using a rule-based method, including regular expression with negation detection. Datasets were randomly split into 80% for development and 20% for testing to evaluate the model's extraction performance. The performance of the NLP pipeline was evaluated by using recall, precision, accuracy, and the F1 score. In evaluating NLP model performances, four parameters—recall, precision, accuracy, and F1 score—were greater than 0.9 for all 23 radiologic findings. These four scores were 1.0 for 10 radiologic findings (listhesis, annular fissure, disc bulge, disc extrusion, disc protrusion, endplate edema or Type 1 Modic change, lateral recess stenosis, Schmorl's node, osteophyte, and any stenosis). In the seven potentially clinically important radiologic findings, the F1 score ranged from 0.9882 to 1.0. In this study, a rule-based NLP system identifying 23 findings related to LBP from X-ray, CT, and MRI reports was developed, and it presented good performance in regards to the four scoring parameters.

Keywords: natural language processing; low back pain; lumbar spine imaging

Citation: Kim, Y.; Song, C.; Song, G.; Kim, S.B.; Han, H.-W.; Han, I. Using Natural Language Processing to Identify Low Back Pain in Imaging Reports. *Appl. Sci.* **2022**, *12*, 12521. https://doi.org/10.3390/app122412521

Academic Editors: Valentino Santucci and Habib Hamam

Received: 17 October 2022
Accepted: 6 December 2022
Published: 7 December 2022

1. Introduction

Low back pain (LBP) is defined as pain and discomfort localized under the ribs and above the inferior gluteal folds, with or without leg pain [1]. A total of 70–80% of adults experience LBP in some form during their lives [2]. More than 85% of people under 45 years of age experience at least one LBP symptom that requires medicine or interventional treatment [3]. LBP can be classified as acute or chronic, depending on its onset. Acute LBP usually occurs suddenly, and if the pain persists for more than 3 months, it can be called chronic LBP [4]. LBP is one of the most common causes of hospital visits and the second-leading cause of sick leave [5]. Because of its high direct and indirect costs, the health, social, and economic impacts on individuals, families, and society are significant [6]. Particularly, between 5% and 10% of chronic low back pain (CLBP) cases require high costs and long-term care [7]. It is important to develop tools to help patients with recently developed back pain accurately predict whether persistent pain will occur [8].

The most common methods to evaluate chronic LBP are X-radiation (X-ray) examinations, computed tomography (CT), and magnetic resonance imaging (MRI) [9]. The radiological findings identified from reviewing those images are translated by clinicians and transferred to a radiology report. A radiology report is a formal interpretation of a radiological examination and contains radiological findings related to the clinical diagnosis and treatment decisions [10,11]. However, many radiologic reports use a free-text form language and remain unstructured, requiring a manual review to harvest clinical information. Manual extraction of information is labor-intensive and impractical for large-scale research. It is important to create a pipeline for extracting clinical information from the radiology reports.

Natural language processing (NLP) is defined as understanding, analyzing, and extracting meaningful information from text (natural language) by computer science [12]. NLP, a computer technology specializing in text processing, can be used to extract critical information from unstructured text in electronic health records [13]. The NLP system is classified into three approaches: the rule-based approach, the machine learning-based approach, and the hybrid types approach [12]. For example, a Bidirectional Encoder Representations from Transformers (BERT)-based NLP pipeline (a state-of-the-art deep learning model for language processing) was used to identify important findings in intensive care chest radiograph reports [14]. Emilien et al. used the NLP system developed by the BERT model. The free-text form records in the University Hospital of Amiens-Picardy in France was used to develop the BERT model. This BERT model learned contextual embeddings. In this research, the utilization of the BERT method in the medical care field was discussed [15]. Tan et al. developed a rule-based NLP system and a machine learning-based NLP system to identify lumbar spine imaging findings related to LBP and compared their performance [16].

In previous studies, there were some points that need to be improved. First, while the machine learning approach has been actively studied, the rule-based approach has not been studied much in the NLP system for extracting words associated with LBP. The set-up cost of the rule-based approach is lower than that of the machine learning approach. The training dataset size of the rule-based approach is also smaller than that of the machine learning approach. In previous studies, the rule-based approach presented high specificity, but moderate sensitivity [16]. This study developed an improved rule-based NLP system to that used in the previous research. Second, there is a need to develop the NLP system for extracting words associated with LBP in CT radiology reports. In previous studies, the NLP system was not trained and tested in the CT radiology reports [16–21]. In this study, the utilization of the NLP system was expanded to the CT radiology reports for extracting words associated with LBP.

This study aimed to develop an NLP system to recognize radiologic findings associated with LBP in X-ray, CT, and MRI radiologic reports. A rule-based approach of the NLP system was trained and evaluated with radiologic reports from patients who sought medical care for LBP. By developing an NLP system, it is possible to analyze the free-text from radiology reports unique to the medical institution. Unstructured clinical information from the radiology reports was processed into a structured data form. The structured form clinical data will be used for the data-drive research of LBP. This study laid the foundation for extracting the clinical data from free-text from radiology reports and using it in various clinical studies.

2. Materials and Methods

2.1. Dataset

This retrospective study of lumbar spine imaging reports enrolled patients who visited CHA University Bundang CHA Medical Center between January 2011 and December 2020. A dataset was assembled from 18,640 reports, and a sample of 300 X-ray, 300 CT, and 300 MRI reports was randomly extracted. With reference to the method of the NLP system for extracting words associated with LBP in previous research, 5% of these words were extracted from the original dataset and used for this study [17]. The dataset was selected randomly. The dataset was split into training and test datasets in an 8:2 ratio, maintaining the ratios of the three different modalities. The training dataset was used for developing the NLP system, and the test dataset was used to validate the information extraction performance. A consensus was reached through a discussion for any cases of discrepancies. This study was approved by the Institutional Review Board of the Bundang CHA Medical Center (IRB No. 2021-04-212).

2.2. Inclusion Criteria

The extracting of clinical information by the NLP system was carried out according to the inclusion criteria. The criteria were composed of radiologic findings in the form of words about spine pathology, structural problems of the spine, and spinal disease. A total of 23 radiologic findings known to be related to LBP from previous studies were established, and each report was annotated by two physicians specializing in LBP diagnosis [18,22] (Table 1). In the 23 radiologic findings, 7 items—disc extrusion, endplate edema or Type 1 Modic, any stenosis, central stenosis, foraminal stenosis, nerve root displaced/compressed and lateral recess stenosis represent the potentially clinically important findings [16].

Table 1. 23 Radiologic findings of the study.

Radiologic Findings
Listhesis
Scoliosis
Fracture
Spondylosis
Annular fissure
Disc bulge
Disc degeneration
Disc extrusion
Disc herniation
Disc protrusion
End plate edema or type 1 Modic change
Facet hypertrophy
Central stenosis
Foraminal stenosis
Nerve root displaced/compressed
Lateral recess stenosis
Spondylolysis
Schmorl's node
Osteophyte
Disc space narrowing
Any stenosis
Disc sequestration
Intradiscal vacuum

2.3. NLP System

This NLP pipeline was developed using Python (3.8.10). Section and sentence segmentation from the radiology reports was performed using a rule-based approach, using regular expression with negation detection. A regular expression is a set of characters specialized for analyzing patterns in text.

The "re" module is one of the Python libraries used to implement regular expressions in Python. By using "re", syntactic analysis was performed to develop clinical information extraction rules. The training dataset was analyzed to construct a customized dictionary of terms associate with the 23 predefined LBP features. From the X-ray, CT, and MRI radiology reports, an NLP pipeline was developed to extract only terms relevant to LBP diagnosis. The pipeline extracted LBP information in three steps: text preprocessing, concept mapping, and summarizing.

First, in the text preprocessing step, tokenization was performed. The NLP refined the unstructured raw text to remove unexpected white spaces and punctuation and filtered out sentences unrelated to LBP. In this step, section segmentation, sentence segmentation, and normalization is performed. Second, LBP concepts were mapped by referring to the constructed dictionary and linguistic rules. Finally, in the summarizing step, the mapped concepts were summarized in a predefined format (Figure 1).

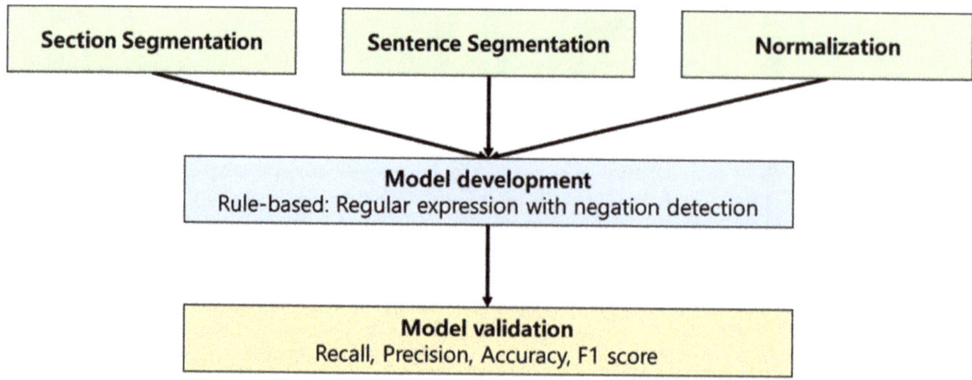

Figure 1. Summary of the natural language processing pipeline.

The whole process was repeated until the performance of this NLP system was improved. The k-fold cross validation was used for the NLP model validation. The NLP pipeline version with the best performance was validated by the test dataset. Therefore, the words associated with LBP were extracted from the X-ray, CT, and MRI reports, using the patterned regular expressions. The extracted words are expressed in a matrix. The source code of the NLP system can be found at https://github.com/YJK96/Low-Back-Pain-NLP.git, which was accessed on 28 November 2022.

2.4. Statistical Analysis

The chi-square test was used to compare ratios, and the t-test was used to compare quantitative variables. The performance of the NLP system was measured using recall, precision, accuracy, and the F1 score.

The recall is defined as $\frac{Number\ of\ true\ positives}{Number\ of\ true\ positives + Number\ of\ false\ negatives}$.

The precision is defined as $\frac{Number\ of\ true\ positives}{Number\ of\ true\ positives + Number\ of\ false\ positives}$.

The accuracy is defined as

$$\frac{Number\ of\ true\ positives + Number\ of\ true\ negatives}{Number\ of\ true\ positives + Number\ of\ true\ negatives + Number\ of\ false\ positives + Number\ of\ false\ netgatives} \quad (1)$$

The F1 score is defined as $\frac{2 \times recall \times precision}{prcision + recall}$ and is used to estimate performance. This score is the harmonic mean of precision and recall, and it is considered more useful than accuracy because of the prevalence of class imbalance in text classification [23].

3. Results

3.1. Dataset Characteristics

The dataset (n = 900) was divided by the training datasets (n = 720) and the test datasets (n = 120). The median age, standard deviation of patients, and the proportion of male or female patients for each dataset were calculated.

In the dataset evaluation, the training and test dataset was classified by 23 radiologic findings. The prevalence of radiologic findings ranged from 1.4% (disc sequestration) to 56.3% (disc bulge) in the training set and from 0% (disc herniation, facet hypertrophy) to 53.9% (disc bulge) in the test set. In both the training and test sets, disc bulge was the most common finding (56.3% and 53.9%, respectively). Spondylosis was the second-most common finding (39.7% and 42.8%, respectively), followed by listhesis as the third-most common finding (36% and 33.3%, respectively) (Table 2). Between the training and test datasets, 4/23 radiologic findings (facet hypertrophy, lateral recess stenosis, any stenosis, disc sequestration) show a statistically significant difference. (p-value < 0.05) (Figure 2).

Table 2. Characteristics of training and testing datasets for the development of the natural language processing pipeline.

Characteristics	Training (N = 720)	Testing (N = 180)	p-Value
Mean age (SD)	62.5 (15.9)	61.3 (15.4)	0.402
Sex (n)			0.066
Male	291 (40.4)	67 (37.2)	
Female	429 (59.6)	113 (62.8)	
Listhesis (%)	259 (36.0)	60 (33.3)	0.508
Scoliosis (%)	147 (20.4)	34 (18.9)	0.647
Fracture (%)	130 (18.1)	37 (20.6)	0.440
Spondylosis (%)	286 (39.7)	77 (42.8)	0.455
Annular fissure (%)	58 (8.1)	10 (5.6)	0.256
Disc bulge (%)	405 (56.3)	97 (53.9)	0.568
Disc degeneration (%)	17 (2.4)	2 (1.1)	0.297
Disc extrusion (%)	82 (11.4)	25 (13.9)	0.354
Disc herniation (%)	15 (2.1)	0 (0.0)	0.051
Disc protrusion (%)	152 (21.1)	43 (23.9)	0.418
Endplate edema or type 1 Modic change (%)	22 (3.1)	10 (5.6)	0.105
Facet hypertrophy (%)	19 (2.6)	0 (0.0)	0.028
Central stenosis (%)	215 (29.9)	55 (30.6)	0.856
Foraminal stenosis (%)	218 (30.3)	51 (28.3)	0.610
Nerve root displaced/compressed (%)	54 (7.5)	17 (9.4)	0.387
Lateral recess stenosis (%)	18 (2.5)	10 (5.6)	0.035
Spondylolysis (%)	18 (2.5)	5 (2.8)	0.833
Schmorl's node (%)	46 (6.4)	17 (9.4)	0.151
Osteophyte (%)	42 (5.8)	11 (6.1)	0.887
Disc space narrowing (%)	181 (25.1)	51 (28.3)	0.381
Any stenosis (%)	29 (4.0)	1 (0.6)	0.020
Disc sequestration (%)	10 (1.4)	8 (4.4)	0.009
Intradiscal vacuum (%)	62 (8.6)	19 (10.6)	0.415

This table shows the number of diseases diagnosed in 720 training set and 180 test set reports.

3.2. NLP System Performances

To evaluate the best performances of the NLP system, the NLP model was trained and datasets tested by calculation of the recall, precision, accuracy, and the F1 score. The raw data of the NLP system is available at https://github.com/YJK96/Low-Back-Pain-NLP.git, which was accessed on 28 November 2022.

In the test datasets (n = 120) of the rule-based NLP model, all 23 radiologic findings had scores of more than 0.9 for recall, precision, accuracy, and the F1 score. These four scores were 1.0 for 10 radiologic findings—listhesis, annular fissure, disc bulge, disc extrusion, disc protrusion, endplate edema or type 1 Modic change, lateral recess stenosis, Schmorl's node, osteophyte, and any stenosis. The lowest F1 score was 0.9802 for facet hypertrophy.

The highest recall score was 1.0 for 14 radiologic findings—listhesis, fracture, annular fissure, disc bulge, disc extrusion, disc herniation, disc protrusion, endplate edema or type 1 Modic, facet hypertrophy, lateral recess stenosis, spondylolysis, Schmorl's node, osteophyte and any stenosis. The lowest recall score was 0.9840 for central stenosis.

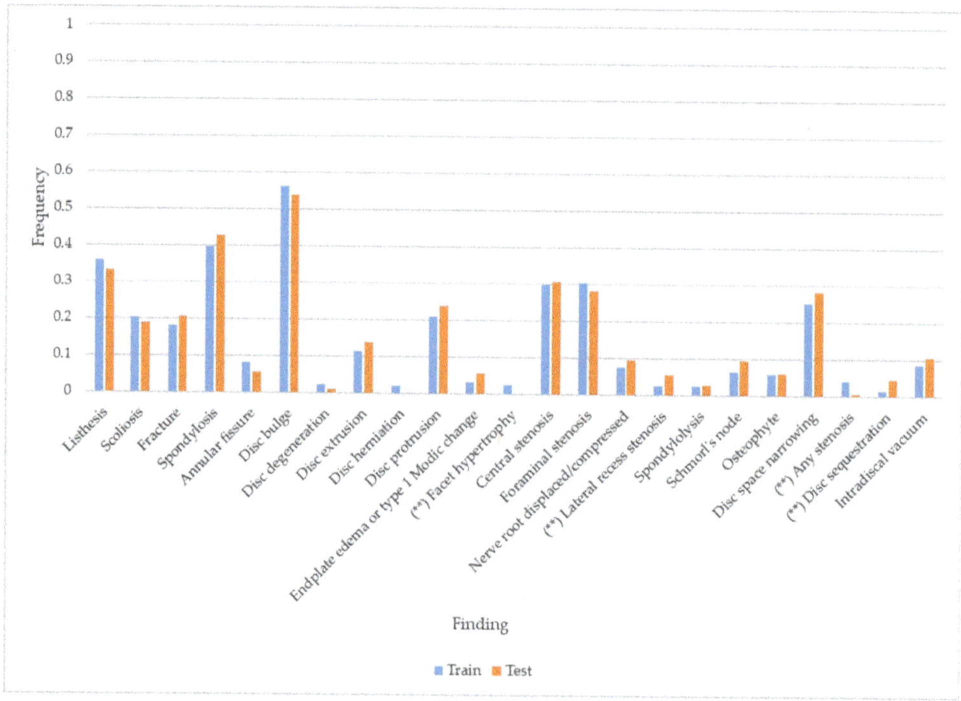

Figure 2. Frequency of the radiologic findings between the training and test datasets. (**) indicates a statistically significant difference (p-value < 0.05).

The highest precision score was 1.0 for 16 radiologic findings—listhesis, scoliosis, spondylosis, annular fissure, disc bulge, disc extrusion, disc protrusion, endplate edema or type 1 Modic, central stenosis, nerve root displaced/compressed, lateral recess stenosis, Schmorl's node, osteophyte, disc space narrowing, any stenosis, and disc sequestration. The lowest precision score was 0.9611 for facet hypertrophy. The highest accuracy score was 1.0 for 10 radiologic findings—listhesis, annular fissure, disc bulge, disc extrusion, disc protrusion, endplate edema or type 1 Modic, lateral recess stenosis, Schmorl's node, osteophyte, and any stenosis. The lowest accuracy score was 0.9611 for facet hypertrophy (Table 3).

The F1 score for the potentially clinically important 7 radiologic findings is as follows: 1.0 for disc extrusion, 1.0 for endplate edema or Type 1 Modic, 1.0 for any stenosis, 0.9919 for central stenosis, 0.9882 for foraminal stenosis, 0.9969 for nerve root displaced/compressed, and 1.0 for lateral recess stenosis.

Table 3. Performance of the natural language processing pipeline in the test dataset (N = 180).

Indicators	Recall	Precision	Accuracy	F1 Score
Listhesis	1	1	1	1
Scoliosis	0.9932	1	0.9944	0.9966
Fracture	1	0.986	0.9889	0.993
Spondylosis	0.9903	1	0.9944	0.9951
Annular fissure	1	1	1	1
Disc bulge	1	1	1	1
Disc degeneration	0.9944	0.9944	0.9889	0.9944
Disc extrusion	1	1	1	1
Disc herniation	1	0.9889	0.9889	0.9944
Disc protrusion	1	1	1	1
Endplate edema or Type 1 Modic change	1	1	1	1
Facet hypertrophy	1	0.9611	0.9611	0.9802
Central stenosis	0.984	1	0.9889	0.9919
Foraminal stenosis	0.9921	0.9844	0.9833	0.9882
Nerve root displaced/compressed	0.9939	1	0.9944	0.9969
Lateral recess stenosis	1	1	1	1
Spondylolysis	1	0.9886	0.9889	0.9943
Schmorl's node	1	1	1	1
Osteophyte	1	1	1	1
Disc space narrowing	0.9922	1	0.9944	0.9961
Any stenosis	1	1	1	1
Disc sequestration	0.9942	1	0.9944	0.9971
Intradiscal vacuum	0.9874	0.9874	0.9778	0.9874

4. Discussion

The purpose of study was the development of the NLP system for extracting the clinical findings, which are associated with LPB on X-ray, CT and MRI radiologic reports. For NLP system modeling, the training datasets (n = 720; 240 X-ray, 240 CT. and 240 MRI) and the test datasets (n = 120; 40 X-ray, 40 CT, and 40 MRI) were extracted from 18,640 radiologic reports. Between the training and test datasets, 19/23 radiologic findings show no statistically significant differences.

In this study, we developed a rule-based NLP pipeline for identifying 23 radiologic findings in X-ray, CT, and MRI reports. In detail, regular expression and negation detection were used for modeling. In assessing the performance of this NLP system, there are four parameters for validation—recall, precision, accuracy, and F1 score. This NLP system presented accuracy more than 0.9611. To correct the effects of dataset bias, the NLP system was evaluated using the F1 score. The F1 score is the most important thing among the parameters for NLP system evaluation. The F1 score is an indicator that reflects recall and precision. In this NLP system, all radiologic findings had an F1 score of 0.9802 or higher. Recall means the sensitivity of the statistics. The developed NLP system had a sensitivity of at least 0.984. Compared to the previous rule-based NLP system [16], the sensitivity of the rule-based NLP pipeline was improved.

From the previous study, the potentially clinically important radiologic findings were defined—Disc extrusion, endplate edema or Type 1 Modic, any stenosis, central stenosis, foraminal stenosis, nerve root displaced/compressed, and lateral recess stenosis [16]. In this study, the 7 potentially clinically important radiologic findings had an F1 score of 0.9882 to 1.0. Moreover, the developed NLP system presented a higher sensitivity of potentially clinically important radiological findings than did the previous rule-based NLP system [16]. In this study, a unique NLP system was developed for extracting the clinical findings associated with LBP pain from the radiologic reports of CHA University Bundang CHA Medical Center.

The imaging findings associated with LBP are not explicitly coded in the medical database that is part of the electronic health record. An NLP system automatically identifies such findings from free-text radiology reports, reducing the burden of manual extraction [16]. In this study, a rule-based NLP system was developed and validated for identifying radiologic findings related to LBP in radiology reports. Radiology reports contain a substantial amount of content within electronic health records and can be used for communication and documentation of the imaging. The same radiologic findings can be expressed in different words in radiology reports, and many radiologic findings remain unstructured in medical databases. A trained person can perform the information extraction task, but manual extraction is labor-intensive, costly, and time-consuming. An NLP system can be used to select useful information from unstructured text in electronic medical records, reducing the burden of manual extraction [13].

Numerous NLP systems now exhibit similar accuracy to humans and have already been used for radiologic reports. Knirsch et al. compared an NLP system with specialist review for chest reports [24]. The chest reports of patients with a positive culture and suspected to have tuberculosis were identified. The focus was on identifying six keywords in the chest reports, and a consensus of 89–92% was gained. Chapman et al., developed an NLP system for disease chronicity, certainty, presence, and examination technical quality [25]. The NLP system showed high sensitivity (86–98%) and high specificity (89–93%), but did not exhibit such high results for chronicity (60% and 99%). Jujjavarapu et al. compared different preprocessing and featurization of the NLP pipeline for classifying radiologic findings associated with LBP from X-ray and MRI radiologic reports. In that study, if the NLP pipeline was developed in the same system (e.g., healthcare institution, EMR system), N-grams was the preferred NLP method. If NLP pipelines were developed in multiple systems, document embeddings were considered the best NLP method [17]. Travis Caton et al., investigated the relationship between the radiologic findings of degenerative spinal stenosis on lumbar MRIs (LMRI) and patient characteristics (e.g., age, sex) using rule-based NLP systems. The NLP system identified the patterns of lumbar spine degeneration [19]. Caton et al. investigated lumbar spine MRIs (LMIR) to identify a performance metric for measuring global severity of lumbar degenerative disease (LSDD) [20]. Huhdanpaa et al. developed a rule-based NLP system to identify the patients with Type 1 Modic endplate changes from lumbar MRI reports [18]. Lewandrowski et al. developed deep learning neural network modes to identify features from MRI digital imaging and communications in medicine (DICOM) datasets to produce automatic MRI reports [26]. Galbusera et al. developed BERT-based NLP system to generate annotations for radiographic images from lumbar X-ray reports [21]. In many previous studies, there were no NLP systems for classification or generation annotations from CT radiologic reports.

Some previous studies used clinical notes or electronic health records to assist in clinical diagnosis or treatment decisions for LBP patients. Miotto et al. developed a convolutional neural network model for classifying acute LBP episodes from free-text clinical notes [27]. Walsh et al. developed a support vector machine model for identifying the axial spondyloarthritis (SpA) concepts from electro medical records [28]. Additioanlly, Walsh et al. developed an NLP system for identifying the patients with axial SpA from clinical charts [29]. Zhao et al. developed the NLP algorithms to classify the axial SpA patients from the free text data of the electronic health records [30]. Particularly, the NLP system was applied to identify the electronic medical record or operation notes during surgery or post-operation. Ehresman et al. developed the NLP system for identifying incidental durotomies from intra-operative electronic health records [31]. Karahade et al. developed the NLP system for identifying incidental durotomies from the free-text operation notes [32]. Additionally, a few previous studies identified the complications of spinal surgery from operative notes or electronic medical records. Karhade et al. developed the machine learning algorithms for the prediction of intra-operative vascular injury (VI) and the NLP systems for identifying VI from free-text operative notes [33]. Karhade et al. developed the NLP algorithms for identifying the post-operative wound infection that requires reoperation after lumbar discectomy from free-text operation notes [34]. Karhade et al. investigated the NLP algorithms for predicting the 90-day unplanned readmission of the lumbar spine fusion patients from free-text notes during the hospitalization period [35]. Dantes et al. developed an NLP system based on IDEAL-X to identify venous thromboembolism from electronic medical records [36]. The free-text type clinical note, operation notes, and electronic health records could be used for developing NLP algorithms for identifying or predicting specific spinal diseases or complications of spinal surgery (Table 4) [12].

Recently, NLP system research on clinical reports written using official language has been conducted globally. Emilien et al. developed the NLP system for learning contextual embeddings from free-text form clinical records in France [15]. Kim Y et al. developed an NLP for identifying a Korean medical corpus using BERT models [37]. Dahl et al. developed an NLP system for classifying Norwegian pediatric CT radiology reports. A bidirectional recurrent neural network model, a convolutional neural network model, and a support vector machine model were used for training and testing the NLP system [38]. Fink et al. developed an NLP system for identifying the oncologic outcomes from structured oncology reports created in the German language [39].

In previous research, machine learning- or deep learning-based NLP system predicted multiple findings using the same datasets. Moreover, the machine learning-based NLP system was more scalable than the rule-based model. However, machine learning- or deep learning-based NLP systems required larger datasets and a higher set-up cost than rule-based NLP systems [16].

In this study, the NLP system is developed using the rule-based approach. Clinicians determined and set the regular expressions using "re." The various patterns of the free-text form clinical reports could be defined as regular expressions. Because rules are set by the user, and the rule-based method is easier to debug and has higher precision that what can be expected using the machine learning-based method. Several limitations of the previous studies were overcome. First, this NLP system was trained with relatively smaller datasets (n = 720) than those used in previous studies. Second, the NLP system showed improved sensitivity (from 0.984 to 1), which has been pointed out as low in previous rule-based NLP systems [16]. Third, the utilization of the NLP system was extended to the CT radiology reports for extracting words associated with LBP.

There are several limitations of this study. First, the radiologic findings were collected from one medical center. The pipeline may not properly process reports with different styles of reporting or expressions. In future works, numerous radiology reports with different styles of reporting should be obtained, as well as datasets from multiple institutions. Second, only the rule-based method was used for developing the NLP model. In future work, other featurization methods (e.g., N-grams, document embeddings) should be used and compared to each other. Despite these limitations, the feasibility of the NLP system to identify lumbar imaging results associated with LBP from sampled radiographic reports was confirmed.

In the future, the NLP system will be used to identify lumbar imaging results among subjects without LBP, to predict whether LBP will become chronic, and to investigate the accuracy of predictions of chronic LBP persistence by investigating these predictions along with other test results, such as hematological tests. Additionally, the transformer models (i.e., BERT model) can be used for further studies. The patient datasets with LBP and without LPB will be learned in the BERT model. Through this method, we are expected to discover the sentences or words with a high correlation with LBP. More keywords, in addition to the 23 radiologic findings used in the current study, will be found.

Table 4. Classification of previous studies to develop NLP systems regarding LBP pain, spinal disease, and complications of spinal surgery.

Study	Topic (Findings)	Source	Year
	Datasets from radiological image reports		
Tan et al. [16]	Radiologic findings associated with LBP.	Lumbar MRI reports and X-ray reports	2018
Jujjavarapu et al. [17]	Radiologic findings associated with LBP.	Lumbar MRI reports and X-ray reports	2022
Travis Caton et al. [19]	Relationship between radiologic findings of degenerative spinal stenosis on lumbar MRI (LMRI) and patient characteristics.	Lumbar MRI reports	2021
Caton et al. [20]	Performance metric for measuring global severity of lumbar degenerative disease (LSDD).	Lumbar MRI reports	2021
Huhdanpaa et al. [18]	Patients with Type 1 Modic endplate changes.	Lumbar MRI reports	2018
Lewandrowski et al. [26]	Producing automatic MRI reports.	MRI DICOM datasets	2020
Galbusera et al. [21]	Generating annotations for radiographic images.	Lumbar X-ray reports	2021
	Datasets from free-text type reports		
Miotto et al. [27]	Classification of acute LBP episode.	Free-text clinical notes	2020
Walsh et al. [28]	Identifying the axial SpA concepts.	Electronic medical records	2017
Walsh et al. [29]	Identifying the axial SpA.	Clinical chart database	2020
Zhao et al. [30]	Classification of the axial SpA patients.	Electronic health records	2020
Ehresman et al. [31]	Identifying incidental durotomy.	Intra-operative electronic health records	2020
Karhade et al. [32]	Identifying incidental durotomy.	Free-text operation notes	2020
Karhade et al. [33]	Prediction of intra-operative vascular injury (VI) and identifying VI.	Free-text operation notes	2021
Karhade et al. [34]	Identifying the post-operative wound infection that requires reoperation after lumbar discectomy.	Free-text operation notes	2020
Karhade et al. [35]	Predicting 90-day unplanned readmission of the lumbar spine fusion patients.	Free-text notes during hospitalization period	2022
Dantes et al. [36]	Venous thromboembolism.	Electronic medical records	2018

5. Conclusions

A rule-based NLP system was developed to identify 23 findings associated with LBP from X-ray, CT, and MRI radiology reports sampled from CHA University Bundang CHA

Medical Center. This rule-based NLP system presented good performance. In particular, the potentially clinically important seven radiologic findings exhibited high F1 scores.

Through research, an NLP system was developed to identify and extract the necessary terms from the free-text form clinical data used in medical institutions. Through this study, an NLP system for extracting words associated with LBP from not only X-ray and MRI reports, but also CT reports, was developed.

The utilization of this NLP system will be expanded to extract data needed for medical research or data that becomes a marker for diagnosis or treatment from free-text form clinical data of the medical institution. This NLP system has potential for use in clinical diagnosis, treatment decisions, and large-scale research.

In future studies, the criteria of radiologic findings will be subdivided according to a lesion location. Additionally, results of each patient's complete blood count (CBC) test and blood chemistry test should be used for the prediction of chronic LBP persistency, along with the radiologic findings from the rule-based NLP system. Moreover, the correlation between the degree of pain and the radiologic findings from the NLP system will be investigated.

Author Contributions: Conceptualization and methodology; H.-W.H. and I.H., writing; Y.K., S.B.K., G.S. (Gyuseon Song) and C.S. (Chanyoung Song); data acquisition; Y.K., S.B.K. and G.S. (Gyuseon Song). All authors have read and agreed to the published version of the manuscript.

Funding: This research was supported by the Korean Health Technology Research and Development Project, the Ministry for Health and Welfare Affairs (HR16C0002 and HH21C0027), and the Institute of Information and Communications Technology Planning and Evaluation (IITP) grant funded by the Korea government (MSIT) (No. 2019-0-00224, AIM: AI based Next-Generation Security Information Event Management Methodology for Cognitive Intelligence and Secure-Open Framework).

Institutional Review Board Statement: This study was approved by the Institutional Review Board of the Bundang CHA Medical Center (IRB No. 2021-04-212).

Informed Consent Statement: Not applicable.

Data Availability Statement: The data presented in this study are openly available in [repository name e.g., FigShare] at [doi], reference number [reference number].

Conflicts of Interest: The authors declare no conflict of interest.

References

1. Andersson, G.B. Epidemiology of low back pain. *Acta Orthop. Scand. Suppl.* **1998**, 281, 28–31. [CrossRef] [PubMed]
2. Deyo, R.A.; Cherkin, D.; Conrad, D.; Volinn, E. Cost, controversy, crisis: Low back pain and the health of the public. *Annu. Rev. Public Health* **1991**, 12, 141–156. [CrossRef] [PubMed]
3. Otluoğlu, G.D.; Konya, D.; Toktas, Z.O. The Influence of Mechanic Factors in Disc Degeneration Disease as a Determinant for Surgical Indication. *Neurospine* **2020**, 17, 215–220. [CrossRef] [PubMed]
4. Atlas, S.J.; Deyo, R.A. Evaluating and managing acute low back pain in the primary care setting. *J. Gen. Intern. Med.* **2001**, 16, 120–131. [CrossRef]
5. Deyo, R.A.; Weinstein, J.N. Low back pain. *N. Engl. J. Med.* **2001**, 344, 363–370. [CrossRef]
6. Dionne, C.E.; Dunn, K.M.; Croft, P.R. Does back pain prevalence really decrease with increasing age? A systematic review. *Age Ageing* **2006**, 35, 229–234. [CrossRef]
7. Meucci, R.D.; Fassa, A.G.; Faria, N.M. Prevalence of chronic low back pain: Systematic review. *Rev. Saude Publica* **2015**, 49, 1. [CrossRef]
8. Jarvik, J.J.; Hollingworth, W.; Heagerty, P.; Haynor, D.R.; Deyo, R.A. The Longitudinal Assessment of Imaging and Disability of the Back (LAIDBack) Study: Baseline data. *Spine* **2001**, 26, 1158–1166. [CrossRef]
9. Li, A.L.; Yen, D. Effect of increased MRI and CT scan utilization on clinical decision-making in patients referred to a surgical clinic for back pain. *Can. J. Surg.* **2011**, 54, 128–132. [CrossRef]
10. Birkmeyer, N.J.; Weinstein, J.N.; Tosteson, A.N.; Tosteson, T.D.; Skinner, J.S.; Lurie, J.D.; Deyo, R.; Wennberg, J.E. Design of the Spine Patient outcomes Research Trial (SPORT). *Spine* **2002**, 27, 1361–1372. [CrossRef]
11. Sistrom, C.L.; Langlotz, C.P. A framework for improving radiology reporting. *J. Am. Coll. Radiol.* **2005**, 2, 159–167. [CrossRef] [PubMed]
12. Bacco, L.; Russo, F.; Ambrosio, L.; D'Antoni, F.; Vollero, L.; Vadalà, G.; Dell'Orletta, F.; Merone, M.; Papalia, R.; Denaro, V. Natural language processing in low back pain and spine diseases: A systematic review. *Front. Surg.* **2022**, 9, 957085. [CrossRef] [PubMed]

13. Cai, T.; Giannopoulos, A.A.; Yu, S.; Kelil, T.; Ripley, B.; Kumamaru, K.K.; Rybicki, F.J.; Mitsouras, D. Natural Language Processing Technologies in Radiology Research and Clinical Applications. *Radiographics* **2016**, *36*, 176–191. [CrossRef]
14. Bressem, K.K.; Adams, L.C.; Gaudin, R.A.; Tröltzsch, D.; Hamm, B.; Makowski, M.R.; Schüle, C.Y.; Vahldiek, J.L.; Niehues, S.M. Highly accurate classification of chest radiographic reports using a deep learning natural language model pre-trained on 3.8 million text reports. *Bioinformatics* **2021**, *36*, 5255–5261. [CrossRef] [PubMed]
15. Arnaud, E.; Elbattah, M.; Gignon, M.; Dequen, G. Learning Embeddings from Free-text Triage Notes using Pretrained Transformer Models. In Proceedings of the 15th International Joint Conference on Biomedical Engineering Systems and Technologies—Vol 5: Healthinf, Lisbonne, Portugal, 9 February 2022; pp. 835–841.
16. Tan, W.K.; Hassanpour, S.; Heagerty, P.J.; Rundell, S.D.; Suri, P.; Huhdanpaa, H.T.; James, K.; Carrell, D.S.; Langlotz, C.P.; Organ, N.L.; et al. Comparison of Natural Language Processing Rules-based and Machine-learning Systems to Identify Lumbar Spine Imaging Findings Related to Low Back Pain. *Acad. Radiol.* **2018**, *25*, 1422–1432. [CrossRef]
17. Jujjavarapu, C.; Pejaver, V.; Cohen, T.A.; Mooney, S.D.; Heagerty, P.J.; Jarvik, J.G. A Comparison of Natural Language Processing Methods for the Classification of Lumbar Spine Imaging Findings Related to Lower Back Pain. *Acad. Radiol.* **2022**, *29* (Suppl S3), S188–S200. [CrossRef]
18. Huhdanpaa, H.T.; Tan, W.K.; Rundell, S.D.; Suri, P.; Chokshi, F.H.; Comstock, B.A.; Heagerty, P.J.; James, K.T.; Avins, A.L.; Nedeljkovic, S.S.; et al. Using Natural Language Processing of Free-Text Radiology Reports to Identify Type 1 Modic Endplate Changes. *J. Digit. Imaging* **2018**, *31*, 84–90. [CrossRef]
19. Travis Caton, M., Jr.; Wiggins, W.F.; Pomerantz, S.R.; Andriole, K.P. Effects of age and sex on the distribution and symmetry of lumbar spinal and neural foraminal stenosis: A natural language processing analysis of 43,255 lumbar MRI reports. *Neuroradiology* **2021**, *63*, 959–966. [CrossRef]
20. Caton, M.T., Jr.; Wiggins, W.F.; Pomerantz, S.R.; Andriole, K.P. The Composite Severity Score for Lumbar Spine MRI: A Metric of Cumulative Degenerative Disease Predicts Time Spent on Interpretation and Reporting. *J. Digit. Imaging* **2021**, *34*, 811–819. [CrossRef]
21. Galbusera, F.; Cina, A.; Bassani, T.; Panico, M.; Sconfienza, L.M. Automatic Diagnosis of Spinal Disorders on Radiographic Images: Leveraging Existing Unstructured Datasets With Natural Language Processing. *Glob. Spine J.* **2021**, 21925682211026910. [CrossRef]
22. Takahashi, K.; Miyazaki, T.; Ohnari, H.; Takino, T.; Tomita, K. Schmorl's nodes and low-back pain. Analysis of magnetic resonance imaging findings in symptomatic and asymptomatic individuals. *Eur. Spine J.* **1995**, *4*, 56–59. [CrossRef] [PubMed]
23. Syed, K.; Sleeman, W.T.; Hagan, M.; Palta, J.; Kapoor, R.; Ghosh, P. Automatic Incident Triage in Radiation Oncology Incident Learning System. *Healthcare* **2020**, *8*, 272. [CrossRef] [PubMed]
24. Knirsch, C.A.; Jain, N.L.; Pablos-Mendez, A.; Friedman, C.; Hripcsak, G. Respiratory isolation of tuberculosis patients using clinical guidelines and an automated clinical decision support system. *Infect. Control Hosp. Epidemiol.* **1998**, *19*, 94–100. [CrossRef] [PubMed]
25. Chapman, B.E.; Lee, S.; Kang, H.P.; Chapman, W.W. Document-level classification of CT pulmonary angiography reports based on an extension of the ConText algorithm. *J. Biomed. Inform.* **2011**, *44*, 728–737. [CrossRef]
26. Lewandrowsk, I.K.; Muraleedharan, N.; Eddy, S.A.; Sobti, V.; Reece, B.D.; Ramírez León, J.F.; Shah, S. Feasibility of Deep Learning Algorithms for Reporting in Routine Spine Magnetic Resonance Imaging. *Int. J. Spine Surg.* **2020**, *14*, S86–S97. [CrossRef]
27. Miotto, R.; Percha, B.L.; Glicksberg, B.S.; Lee, H.C.; Cruz, L.; Dudley, J.T.; Nabeel, I. Identifying Acute Low Back Pain Episodes in Primary Care Practice From Clinical Notes: Observational Study. *JMIR Med. Inform.* **2020**, *8*, e16878. [CrossRef] [PubMed]
28. Walsh, J.A.; Shao, Y.; Leng, J.; He, T.; Teng, C.C.; Redd, D.; Treitler Zeng, Q.; Burningham, Z.; Clegg, D.O.; Sauer, B.C. Identifying Axial Spondyloarthritis in Electronic Medical Records of US Veterans. *Arthritis Care Res.* **2017**, *69*, 1414–1420. [CrossRef]
29. Walsh, J.A.; Pei, S.; Penmetsa, G.; Hansen, J.L.; Cannon, G.W.; Clegg, D.O.; Sauer, B.C. Identification of Axial Spondyloarthritis Patients in a Large Dataset: The Development and Validation of Novel Methods. *J. Rheumatol.* **2020**, *47*, 42–49. [CrossRef]
30. Zhao, S.S.; Hong, C.; Cai, T.; Xu, C.; Huang, J.; Ermann, J.; Goodson, N.J.; Solomon, D.H.; Cai, T.; Liao, K.P. Incorporating natural language processing to improve classification of axial spondyloarthritis using electronic health records. *Rheumatology* **2020**, *59*, 1059–1065. [CrossRef]
31. Ehresman, J.; Pennington, Z.; Karhade, A.V.; Huq, S.; Medikonda, R.; Schilling, A.; Feghali, J.; Hersh, A.; Ahmed, A.K.; Cottrill, E.; et al. Incidental durotomy: Predictive risk model and external validation of natural language process identification algorithm. *J. Neurosurg. Spine* **2020**, *33*, 342–348. [CrossRef]
32. Karhade, A.V.; Bongers, M.E.R.; Groot, O.Q.; Kazarian, E.R.; Cha, T.D.; Fogel, H.A.; Hershman, S.H.; Tobert, D.G.; Schoenfeld, A.J.; Bono, C.M.; et al. Natural language processing for automated detection of incidental durotomy. *Spine J.* **2020**, *20*, 695–700. [CrossRef] [PubMed]
33. Karhade, A.V.; Bongers, M.E.R.; Groot, O.Q.; Cha, T.D.; Doorly, T.P.; Fogel, H.A.; Hershman, S.H.; Tobert, D.G.; Srivastava, S.D.; Bono, C.M.; et al. Development of machine learning and natural language processing algorithms for preoperative prediction and automated identification of intraoperative vascular injury in anterior lumbar spine surgery. *Spine J.* **2021**, *21*, 1635–1642. [CrossRef] [PubMed]
34. Karhade, A.V.; Bongers, M.E.R.; Groot, O.Q.; Cha, T.D.; Doorly, T.P.; Fogel, H.A.; Hershman, S.H.; Tobert, D.G.; Schoenfeld, A.J.; Kang, J.D.; et al. Can natural language processing provide accurate, automated reporting of wound infection requiring reoperation after lumbar discectomy? *Spine J.* **2020**, *20*, 1602–1609. [CrossRef] [PubMed]

35. Karhade, A.V.; Lavoie-Gagne, O.; Agaronnik, N.; Ghaednia, H.; Collins, A.K.; Shin, D.; Schwab, J.H. Natural language processing for prediction of readmission in posterior lumbar fusion patients: Which free-text notes have the most utility? *Spine J.* **2022**, *22*, 272–277. [CrossRef]
36. Dantes, R.B.; Zheng, S.; Lu, J.J.; Beckman, M.G.; Krishnaswamy, A.; Richardson, L.C.; Chernetsky-Tejedor, S.; Wang, F. Improved Identification of Venous Thromboembolism From Electronic Medical Records Using a Novel Information Extraction Software Platform. *Med. Care* **2018**, *56*, e54–e60. [CrossRef]
37. Kim, Y.; Kim, J.H.; Lee, J.M.; Jang, M.J.; Yum, Y.J.; Kim, S.; Shin, U.; Kim, Y.M.; Joo, H.J.; Song, S. A pre-trained BERT for Korean medical natural language processing. *Sci. Rep.* **2022**, *12*, 13847. [CrossRef]
38. Dahl, F.A.; Rama, T.; Hurlen, P.; Brekke, P.H.; Husby, H.; Gundersen, T.; Nytrø, Ø.; Øvrelid, L. Neural classification of Norwegian radiology reports: Using NLP to detect findings in CT-scans of children. *BMC Med. Inform. Decis. Mak.* **2021**, *21*, 84. [CrossRef]
39. Fink, M.A.; Kades, K.; Bischoff, A.; Moll, M.; Schnell, M.; Küchler, M.; Köhler, G.; Sellner, J.; Heussel, C.P.; Kauczor, H.U.; et al. Deep Learning-based Assessment of Oncologic Outcomes from Natural Language Processing of Structured Radiology Reports. *Radiol. Artif. Intell.* **2022**, *4*, e220055. [CrossRef]

MDPI
St. Alban-Anlage 66
4052 Basel
Switzerland
Tel. +41 61 683 77 34
Fax +41 61 302 89 18
www.mdpi.com

Applied Sciences Editorial Office
E-mail: applsci@mdpi.com
www.mdpi.com/journal/applsci

www.ingramcontent.com/pod-product-compliance
Lightning Source LLC
LaVergne TN
LVHW070643170726
843515LV00004B/310
* 9 7 8 3 0 3 6 5 8 1 9 8 9 *